KRAKEN OF EDEN

Paige,
thanks...

KRAKEN
OF EDEN

by George Moakley

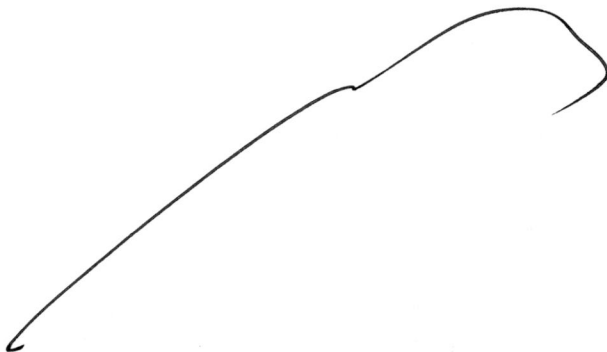

Paperclip Publishing, LLC
Tempe, Arizona

KRAKEN OF EDEN

Copyright © 2022 by George Moakley

Published by: Paperclip Publishing LLC

Editor: Abigail T. Matteson
Cover Illustration: Shaun Cochran
Cover Design and Interior Typography: Hannah Thigpen

Library of Congress Control Number: 2022933898

ISBN: 978-1-7346207-9-5 (paperback)

ISBN: 978-1-7346207-8-8 (hardcover)

ISBN: 979-8-88589-201-8 (eBook)

Printed in Rephen Printing, Co. LTD in Guangzhou and the United States of America

First Printing: 2022

Paperclip Publishing LLC
1840 I Baseline Road Suite A-1
Tempe, AZ 85283

www.paperclippublishing.com

To Diana, for believing in me.
To my kids, for sharing my love of a good monster story.

CONTENTS

Discovery... 1

   The Briefing 3

   Alexandria 17

   Lucky Strike 27

Arrival.. 41

   19 days Before Arrival............................ 43

   10 days Before Arrival............................ 57

   Day 4 65

   Day 5 71

   Day 6 77

   Day 9 83

   Day 10 93

   Day 11 101

   Day 12 105

   Day 13 107

   Day 14 111

   Day 15 119

Day 16 123

Infestation **129**

Day 17, 04:32 Platform Time 131

Day 18, 08:47 Platform Time 139

Day 19, 01:37 Platform Time 153

Day 20, 03:47 Platform Time 161

Day 21, 13:55 Platform Time 167

Day 22, 05:17 Platform Time 179

Day 23, 07:12 Platform Time 205

Battles **233**

Day 24 05:17 Platform Time 235

Day 25 02:27 Platform Time 305

Epilogue **345**

Day 372 347

Day 6127 353

Acknowledgments **357**

About the Author **359**

PART 1:
DISCOVERY

▪THE BRIEFING

Elke Lubandi strode purposefully through the broad, light gray corridor en route to her conference room as junior officers and enlisted crew stepped aside to make way. Since the *Lucky Strike* was nearing its destination, the ship had decelerated enough to allow it to receive a message from their destination. Elke was very much looking forward to hearing a new voice.

The opaque portal cleared briefly as she ducked to enter the conference room. She'd been born in one of the asteroid belt mining colonies in the late twenty-second century, and her early, low gravity years accentuated her genetic predisposition for height. She hadn't seen a portal that didn't require her to bend her head since puberty. Well above average height, she'd exceeded the standards of the Deep Space Academy, but even in those early years they favored space natives and her math aptitude was off the charts. The Academy gladly made the investment in corrective therapies for higher gravity tolerance, fast tracked her career, and figured she'd learn to duck.

Her surname, Lubandi, reflected her east African ancestry, but her dark skin, green eyes, and close-cropped blond hair, a common style for deep space officers and crew, reflected the genetic diversity of the Belt community long before the panmictic programs of the deep space service. Her work uniform, a tan jumpsuit with collar rank insignia and utilitarian boots was clean and neatly pressed, as always. She tolerated informality, especially among the more junior crew members, but held herself to a higher standard as their captain.

Elke entered the conference room to the sound of her senior staff making small talk about the upcoming arrival briefing from the colony that the *Lucky Strike* was currently approaching.

"Someday, you'll have to explain the physics behind this to me again," Sri, her executive officer, was saying. "I just don't understand how we can be exchanging messages simultaneously from this distance. Or why our velocity makes a difference."

Sridharan Rizzo, ninety-one years old with his first touches of gray hair, was an experienced deep space officer, like most of her senior staff. He was of average height for an extrasolar colonist from the twenty-third century. He'd been born on an orbiting platform when his home world's colonial population was in the thousands. By now, the domed cities on the surface were inhabited by millions. After so many crossings at relativistic speeds, he hadn't seen his home world, one of the first extrasolar colonies, in well over two centuries.

Chief Engineer Moira Tam, slender and poised with her fair complexion showing her youth at only seventy years old, leaned back and raised her carefully manicured hands as she responded. "You don't have the math to understand it. Hell, I don't think I completely understand it. At the end of the day, I just know how to fix it when it breaks."

Elke took her seat at the head of the triangular table, various projection devices at its base. The crew looked up and smiled at her as she spoke. "I just care that it works. I keep hoping it'll lead to a breakthrough in faster-than-light travel." She waved a hand as she repeated her often-heard prediction, "one of these days we'll arrive somewhere, thrilled with how closely we approached the speed of light, only to find that someone has finally figured out faster-than-light travel while we were in transit, and managed to get there ahead of us!" The crew laughed lightly.

"Yeah, don't hold your breath, Cap," Security Chief Daniella Wu said good-naturedly, "I think that's still a long way off." She was one of the younger members of senior staff at only sixty-one. Her hair was bright red, and her dark skin showed no signs of aging. Privately, she attributed her youthful appearance to her fierce fitness regimen of martial arts and powerlifting.

There was a brief pause in the conversation, giving Elke the opportunity to greet Chief Medical Officer John Schmitt, 112 years old and built like an endurance athlete, and Operations Chief Nikko Harris, slightly overweight and prematurely gray at seventy-four. They took their places to her right and left respectively.

Elke looked about the room, reflecting once again on how different it was from the cramped, spartan transports she remembered from shuttling around Earth's asteroid belt. Then again, although they were small, those interiors had certainly been colorful. All the rooms on the *Lucky Strike* were spacious and contained amenities to keep the crew comfortable for two to three years at a time. The conference room was no exception, but the walls and furnishings were the same light gray as the rest of the ship's interior. Elke's biggest complaint about deep space vessels was the consistently dull, neutral colors and textures that some long-deceased psychologist had decided struck the best balance of soothing the occupants and hiding the grunge that inevitably accrued during long journeys despite the crew's best cleaning efforts. The only relief from these dull, neutral colors were accent pieces added by colonists and crew and, ironically, the occasional resistant stain from something spilled on what was then the floor.

Chief Astrogator Amar Aggrawal came through the portal, shifting it from translucent to opaque. "Sorry to be late," he said. Amar, also middle aged, was a bit pudgy with poor posture and unkempt blond hair grown just past his ears. He was generally so preoccupied with his calculations that he was frequently tardy, not fastidious with his uniform, and inappropriately casual with his speech. Elke often found herself surprised at how frustrating she found him. Given the duration of interstellar flights, the Deep Space Service, or D.S.S., invested in extensive training to help senior officers stay above interpersonal conflicts and in psychological profiling to avoid combining people who would get on each other's nerves. Still, she couldn't recall an officer she found as irritating as Amar Aggarwal. But, he was among the best astrogators in the fleet, so she'd long since given up on correcting his frustrating habits.

She waved Amar into his chair, looked contemplatively around the room again, and cleared her throat.

"Alright, people, it's finally time. We're nearing our destination and we've received our first communications from the local station. But before we get to that, let's get through our morning agenda. Does anyone have anything pressing we should begin with?" She looked to Sri first.

Sri leaned forward and cupped his hands around his coffee mug before responding. It truly was *his* mug, custom fashioned to fit his hand, a gift from Moira's team for the second birthday he celebrated during their crossing. He loved it, though it was an ungainly thing. It had a base resembling surf, sides of blue skies with light clouds, and a handle reminiscent of a sextant. He had been deeply touched by the gift. Though he'd never been to Earth, he'd been fascinated by her seas since childhood. He dreamed of the great explorers, committing themselves to the unknown, crossing vast, mysterious oceans in search of new worlds, just as he now crossed the vast oceans of space, bound for truly new worlds.

"No ma'am, nothing pressing. All systems are running normally. Moira's team found and addressed the problem with the algae tanks, so that's finally done. Thanks, Moira." He said, nodding toward the Chief Engineer.

"You're welcome," Moira responded primly. She was the consummate engineer: precise, pragmatic, practical. "As we suspected, there was an obstruction in one of the filter hoses. It certainly didn't impact anything significantly, and the algae have almost returned to optimal population levels. No impact on our life support. Nothing else notable to report. Everything's humming along and we're all ready to leave the ship. Looking forward to getting in trouble." She smirked at her little joke, then continued. "We've been receiving fabrication code from the colony as we approach, and we've started programming the printers to make the parts. Some involve raw materials that we've exhausted or never had, but we've been able to produce most of what they're sending us." She looked around the room, "we picked a few of the more interesting ones and started a pool if anyone wants to participate."

Vessels closing on their destinations usually had an abundance of both idle time and spare raw materials, so it was common for the engineering teams that will conduct their refit once they reach port to send code for some of the parts they'll need. Since some of the parts fit components that had been invented since the last time the ship was in port, engineering teams often enjoyed speculating about, and sometimes betting on, what the parts might be for.

"I have no official opinion," Elke said with an entertained expression. "You're each free to make your own choices on that one." She turned to the next crew member. "John?"

John leaned forward and looked around the room. "Nothing notable. One of Dani's guys managed to twist an ankle practicing a martial arts routine. One of Nikko's food workers dropped a pork fabricator on her foot. If she plays in this weekend's table tennis competition, she won't play well. If you're betting, I'd bet on her opponent. She's playing against one of the migrants and I hear he's good. Anyway, nothing interesting, just boring aches and sprains."

"Dani?"

Daniella folded her hands in her lap. "Nothing other than appreciation for our medical team."

"Amar?"

After a moment, Amar looked up from his wrist pad. "Ah, all's on track. Yes. At this rate, we'll, ah, reach port in another two weeks. All in all, we've made great time. Using the revised astrogation parameters I loaded right before we left, we've managed to shave almost six months' real time and about four days' ship time off our initial projections. When we arrive, it'll be the morning of May 24th, 2552 locally."

"That's the week before my 104th," Elke said. "I can celebrate it at the colony. Can't wait."

John shook his head. "At least you'll *get* a birthday. Mine is December 28th and the local year is only 355 days."

"Elke, that's biological time," Nikko said. "What will you be in Earth years? Over 350?"

"Don't remind me, *362*. I know, some of you are biologically older than me, but I've got you all beat chronologically. When we get back, we'll find a pub, and I'll regale you kids with stories about life back in *my* day." Everyone around the table chuckled.

"Nikko, anything to add, or can we listen to the message?" Elke asked when the laughter died down.

"Nothing. I'm just looking forward to some rest and relaxation, and some food from a farm rather than a fabricator. Don't get me wrong, these migrant runs are easier than colonial runs, but the food's better on the colonial runs. I'm betting there's been a lot of growth in the colony since we've been there. What's it been for them? Almost thirty years? Bet it's grown substantially! Should have a space elevator and surface settlement by now. Has to be some new sights to see and pubs to crawl!"

There was a brief buzz in the room as the attending officers discussed the first things they'd do once they reached the colony. After letting them briefly indulge, Elke spoke to the desk, "play the governor's message." At the sound of her voice, the room recovered an air of professionalism as everyone waited for the communique to begin.

The lights dimmed slightly, and a meter-tall holographic projection of a smiling old man came into focus. He stood on the table at the far end where the whole crew could see him. His full brown beard was precisely trimmed against his slim face, and his hair was elaborately set in rows of tight curls draped over his shoulders like thick ribbons. He was dressed in a one-piece jumpsuit with a tailored, brightly hued jacket. It was clear from his appearance that personal grooming and formal fashion had changed while they were out, an unsurprising development since they had been in space for fifteen years. The figure turned slowly right and left, almost seeming to make eye contact with the staff. Despite being a recorded projection, this messaging technology created the feeling of having an active conversation.

"Captain Lubandi and staff, welcome back to Alexandria. I'm Governor Abram Kelty. We have some exciting news and updates for you, now that you have decelerated enough to receive messages.

About five years ago, after you started your return journey, we started receiving fascinating reports from a set of autonomous probes in what we are temporarily referring to as 'Star System 57.' There are eight planets with numerous satellites within this system. The probes deployed drones, and the drones identified a number of orbiting bodies with promising mineralogical deposits and impressive concentrations of rare, precious metals. We're analyzing drone reports to understand mining potential. Nothing unusual there. As you know, almost every system we've explored has had one or two orbiting bodies with some organic material, from disorganized organic chemicals through simple single-celled life forms roughly analogous to Earth bacteria. To date, we've not found any complex cells, let alone multicellular forms. The novel organic chemicals have been of great scientific interest and commercial value. Until now, we've found no signs of complex life."

The crew looked at one another, wide eyed at the implications of what he'd just said. The recording of Governor Kelty paused, then continued.

"For the very first time, drones have detected the presence of a complex ecology, located on the third planet from the sun. Most importantly, the atmosphere contains some thirty-one percent oxygen and there is a great deal of greenery visible on the planet. An oxygen-rich atmosphere simply does not occur without photosynthesis on a scale beyond the single-celled life forms we've seen elsewhere. Our researchers are calling this planet 'Eden.'"

The crew started murmuring excitedly. Lubandi waved her hands to suppress voices in the room so the governor could still be heard. Beside him, a slowly spinning globe appeared. The eye-catching sphere had small white caps at the top and bottom, wisps of white over a blue and green surface with a few patches of brown. Four smaller bodies orbited the planet at a small distance.

"Since you're still far enough away that we can't have an actual conversation, let me anticipate a few questions." His figure turned toward the globe and he raised his hands to point out the features as he referenced them. "There are no signs of intelligent life: no straight lines, no regular shapes, no atmospheric hydrocarbons or other signs of industrial activity, and no detectable structured radio transmissions. This is based on a number of presumptions about what we'd expect to see if there were intelligent life, and those assumptions could be wrong, so we look forward to more in-depth reports from the first colonists. Meanwhile, per protocol, once the drones detected signs of a complex ecosystem, they moved to higher orbit to observe and have taken no samples beyond initial atmospheric scanning. You can see that there are polar ice caps and oceans. We see signs of a number of ecological zones that we're guessing are comparable to Earth's tropical and temperate forests, deserts, and so forth. Temperature ranges are fairly close to Earth's as well, as you might guess from the abundance of liquid water. It's similar in size to Earth, so we're anticipating similar gravity. The planet orbits its sun every 403 days, and one full day is just over twenty-seven hours. The declination angle is 19° and the orbit slightly elliptical, so there should be seasons. Drone observations suggest significantly more tectonic activity than Earth, probably due to the gravitational complexity of the four moons orbiting the planet. As you can see, two of the moons are relatively large, though not quite as large as Earth's moon. Their orbital periods range from twenty-seven to thirty-eight days."

The governor faced forward again as the sphere faded away. "This is the first complex ecosystem we've found," he said earnestly, "and we've hoped for this moment for quite some time. Our probes and their drones are following a long-standing but never-before-used protocol. As soon as they detected key indications of complex life, they prioritized preparations for colonization. There will be no more close-range observation or sampling until after colonists arrive, but the orbiting colonial framework is nearly complete. Antimatter generators that will fuel the *Lucky Strike*'s return journey have already been set up. Which brings us to the next bit of news…"

The governor stepped forward and cleared his throat. "The *Lucky Strike* will be arriving soon, and I'm sure you and your crew are expecting to collect your bonuses and spend the customary year on leave while we refit the ship for its next journey with an alternate crew. However, we are anxious to get colonists to Eden, and the next ship that will be coming through here is more than a year behind you. Captain Lubandi, you're one of the most experienced captains in the fleet. You have top-notch officers and crew, and we're delighted you're in the area. The pool of potential senior officers currently at our colony includes some fine candidates with impressive potential, but none can match the caliber of you and your team. So, we are preparing to conduct an accelerated refit of the *Lucky Strike*, and invite you and as many of your crew as are willing to leave again within three weeks of your arrival here. This is an invitation, not a mandate, although we hope you will all consider this historic opportunity. Also, your shares of the accelerated journey's earnings will be 10% higher than usual."

Captain Lubandi again gestured for silence as her senior staff exploded in excitement.

"You'll be shepherding the first wave of colonists, all of whom have been preparing for the journey for nearly five years. There will be 1,000 of them, as is customary for this type of mission, but their population will be skewed to include far more xenobiologists than usual." The governor's face softened. "As you might imagine, we've had quite a surge of students through the local university's xenobiology program, including some degrees that haven't been offered anywhere except on Earth herself, like oceanography. This is an extraordinary opportunity. From a discovery perspective, well, what could compete with the first alien ecosystem? The scientific potential is unprecedented. With another complex ecosystem to study, perhaps we can begin to understand why such life is so rare. In addition to all the possibilities we associate with any new system, preliminary economic assessments indicate high levels of precious metals and the organic chemical potential is enormous. And I'm sure that you're wondering what we've been wondering: is there a possibility of establishing a colony that doesn't require a dome?"

The governor clapped his hands together as he looked about the table, seeming to look at each crew member in turn, even though the nature of the asynchronous message made that impossible. "So, there you have it. You all have a big choice to make, though I hope it will be an easy choice. We'll be waiting for your response, and, as you get closer, for interactive communications. Safe travels!" The image of the governor faded and disappeared. The room was briefly silent as everyone processed what they had just heard.

The *Lucky Strike* was currently returning from a colonization run, carrying 1,000 new colonists to an already-colonized star system to increase the number of residents as a surface dome and space elevator were built there. They also carried a wealth of mineralogical and organochemical processing pods, and 200 young migrants who would be taking advantage of the Deep Space Service's Panmictic Incentive Program.

The history of humanity's nationalism and racial discrimination had been well known for many years, but the misery came to a head with the global conflicts of the twenty-first century, only to be followed by nationalistic tensions between Earth, the Moon, Mars, and the asteroid belt in the twenty-second century. Given interstellar distances, concerns arose that genetic isolation between colonies could result in the emergence of morphological differences that would compound humanity's unfortunate tendency towards racism, discrimination, and violence. The D.S.S. created the Panmictic Incentive Program, or P.I.P., to prevent such genetic isolation. They monitored the gene pool of each settled world and adjusted migration benefits for 'pippers' to keep the human race as genetically and culturally homogenous as possible. Adding the ability to exchange messages instantly was immensely valuable as well, as it helped maintain a baseline level of cultural consistency across known space.

Young adults willing to relocate to other worlds received a free university education and other enticements, the scale of which increased with distance. It was one thing to spend a couple of years relocating to a neighboring system only ten or fifteen years away from everything one grew up with. It was still possible to message with the folks they left behind and rebuild some sense of community. Some migrants, however, accepted higher incentives for journeys involving decades of real time. With this agreement, often involving financial incentives, vocational retraining, and educational grants for their descendants, came the understanding that they were truly, permanently leaving behind everything they had ever known. These programs were especially appealing to those still Earthbound, wishing to trade draconian restrictions on family size for colonial embracement of population growth.

While Panmixia was successful in diversifying the population of many settled colonies, the program put a constraint on the potential expansion of humanity. The first wave of expansion took humans to star systems within fifteen light-years of Earth. Ongoing improvements to the matter/antimatter drive brought deep space vessels closer and closer to the speed of light. As this reduced the shipboard experience of a fifteen light-year journey to less than three years ship's time, longer journeys became more tolerable and humanity began a second wave of expansion to systems within thirty light-years of Earth. Doubling the diameter of known space had multiplied the number of potential star systems by a factor of eight, leading to the discovery of numerous promising opportunities to start new colonies. But now, the distance between the furthest outposts of humanity were over sixty light-years apart. The ability to keep the interstellar human gene pool reasonably homogenous had already been strained before this development, so a Deep Space Service policy had been implemented to limit further expansion. The only exception to this rule was purely commercial outposts used for mining and harvesting novel organic chemicals. Eden would be one of the few colonies that was more than thirty light-years from Earth. The good news was that humanity now had plenty of room to grow. With this good news came the universal hope that faster-than-light travel would be achieved before any colony needed to discourage population growth.

For the crew of the *Lucky Strike*, each outbound and return mission at relativistic speeds was nearly a six-year investment. They were looking forward to a year's leave to spend their shares of the trip's earnings and enjoy the progress the colony had made in the nearly thirty years since they'd left.

Elke leaned back in her chair and looked around the table. "Thoughts?"

John smiled broadly and slowly shook his head in disbelief. "Are you kidding? I'm in! This is fantastic!"

"Not surprised at all," Elke responded with a smirk. "How about the rest of you?"

Sri exhaled slowly. "I was very much looking forward to my leave and spending some of my pay. I haven't logged as many flights as you, Elke, but I've logged more than most. But this?" He gestured towards the end of the table where the sphere had been spinning. "How do you say no to *this*?"

Moira's facial expression was hard for Elke to read. "I need to think," she said, "you know me. I love this stuff. I'll probably sign on. But I was so looking forward to a break. Any chance we can convince them to give us more than a few weeks? Like Nikko said, they must have a space elevator and dome by now. Even two or three weeks to explore the surface settlement would be welcome. I'm not sure I get the urgency."

Elke nodded, "fair point, I'll see whether there's any flexibility. No promises. Anyone else?" She looked about, giving the crew a little more time to process the new information, then stood. "Here's where we'll leave it. I'll prepare a message for the governor thanking him for the offer. I'll let him know that at least a few of us are ready to sign on. I'll ask for a few days for the rest of you to decide, and about whether we can have a bit more time before shipping out. I would also need to give the governor time to recruit people to take anyone's place who decided not to come along. I doubt that would be hard to do, under the circumstances. But let me reiterate—it is entirely your call. Just let me know what you want to do. For now, let's presume we have a week to think it through. Fair enough?"

Nikko looked up and made eye contact with Elke. "I assume we're to extend the offer to our teams?"

"Yes, please do. Given that this is an initial colony run, we'll be adding staff anyway, so we want to retain as much talent as we can. If you have any concerns about anyone on your team, please tell Sri or me privately." Elke looked around the table as everyone stood. "Anything else? No? Okay. For what it's worth, I'll be signing up. There's *no* way I'm missing this! But I won't think less of anyone that decides they want their leave, okay? You've earned it."

After the others left the room, Elke ducked through the translucent portal into a hallway that led to her quarters. Shift change occurred while they were meeting, so the corridors were lightly trafficked. Junior officers and crew nodded and occasionally smiled as they passed. After a short walk she ducked through the entrance portal of her quarters. The automatic lights came on, triggered by her movement in the space, and her portal shifted from translucent to opaque for privacy.

The suite was roomy and comfortable. The first room upon entering was an anteroom that held a small work table with four chairs, and a kitchenette for preparing drinks and light snacks. Next came the commode and shower, which served as a boundary between the workspace and her bunk room beyond. Above the worktable were shelves displaying an assortment of personal items. Trinkets from her travels, old mining tools and other keepsakes from her youth in the Belt and, from her adolescence at the Academy, a very old pocket computer. Across from the worktable was the large screen of her comms station, which had been displaying a collage of images since she'd entered the suite. They were images of places she'd been and people she'd known. The smiling faces of family, friends, and colleagues glowed from the screen, accompanied by soft, soulful instrumental music that had been popular many years ago when she was a young woman.

Elke watched as an image of an extended family appeared on the screen. Her daughter Emily, a woman of advanced years, stood next to her husband, surrounded by generations of their descendents. Emily had built herself a satisfying life, living out her days in the colony where she was born rather than following her mother and father into the Deep Space Service the way her siblings had. Born at the colony while Elke was governor, Emily had long since passed.

The image grew for about a minute, then faded into the next. More images of Elke's children lazily floated by, followed by their children and grandchildren. One of her sons popped up in his Deep Space Service Uniform; he was still serving, as he hadn't decided to settle down and start a family on some colony. Elke sighed, reliving memories of her children's childhoods, educations, and career paths. Elke had loved being governor, but not nearly as much as she'd loved raising her family. But, as her children moved on and launched their own lives, she realized they didn't need her anymore. After the last child left home, she returned to the Service and stayed in touch as best she could. She had been able to retain some contact with Emily over the years, but given the realities of lives spent at relativistic speeds, she had lost all track of her sons years ago. It had been nearly forty years for Elke since she went back to work, and about 200 for Emily and her family who didn't join the D.S.S. Elke wondered if any of them had any idea who she was.

Elke shook her head to dislodge the memories and focus on work once again. She walked to the comms station, stood on the mark, and said, "record." After waiting a moment for the red light that indicated the machine was recording, she began speaking:

"Governor Kelty, it is a pleasure to make your acquaintance. My officers and I have listened to your message, and we are delighted to learn of this opportunity." She allowed her face to soften as she conveyed her team's enthusiasm at the upcoming mission. "Actually, I don't think 'delighted' even comes close."

Without thinking, she shifted to a more formal posture to make her requests. "I believe most of my crew will opt to make the journey, but I also know that many of them were looking forward to some leave time. I've presumptuously given them a week to make up their minds, which I trust will not create any challenges for our refit timeline. I'm also asking whether we might have a bit more than three weeks. Most of that time would be spent overseeing the refit and getting acquainted with system upgrades in preparation for shipping back out, so I'd like to request some additional time. Even a week or two would give my people a bit of badly needed R&R, and would increase officer and crew retention."

She relaxed back to an at-ease position to finish the communique on a relaxed note. "We'll await your answer. I'm looking forward to meeting you in person." She shifted her gaze to the light on the comm. "End recording, and send."

The message would be received immediately, but it would take time for the governor to see it and respond. Elke walked back through the front portal of her suite and headed toward the bridge.

◾ALEXANDRIA

Elke sat at the colonial bar, nursing a shot of Irish whiskey. She didn't consume very much alcohol, and when she did, she preferred to sip it straight. She held a sip in her mouth a moment, fully enjoying its complex aromas and flavors.

She enjoyed the bar. Even the best deep space vessels didn't indulge in proper pubs, and migrants were typically too young to produce quality moonshine. In front of her, four bartenders each supported a handful of customers. She guessed most of the customers were colonial government, students and faculty from the local Deep Space Academy, and civilians affiliated with the various businesses involved with interplanetary and interstellar travel.

Just a few hours before, she'd welcomed the colonial harbormaster aboard the *Lucky Strike*. Together they worked to oversee migrants boarding shuttle tugs to cross over to the colony, then the crew. Finally, she relinquished control of the *Lucky Strike*, boarded the last shuttle, took her gear to her colonial quarters, and, per very old tradition, headed to the bar.

She looked around. It was a pleasant room in the bottom ring of the rotating colonial drum. This ring featured transparent panels along the wall opposite the bar, offering a stark view of the stormy planet below. Also visible from the bar was the long, thin tether of the space elevator, stretching from the core of the rotating drum down to the colonial dome on the planet's surface.

The bar itself extended about twenty meters in length, with enough open space at either end to perceive the curvature of the floor. It was made of a dark simulated wood with a rich grain. *A rather good simulation.* In truth, Elke had never seen real wood. Earth's depleted forests couldn't possibly provide enough wood for all of the interstellar colonies, but she'd seen her share of simulated products. Though it would be impossible for her to comment on the accuracy of the wood currently in front of her, it was definitely attractive and soothing, which, she decided, was what really mattered.

Elke still wore her uniform, but the collar was open since she was off duty. This was about as informal as she was willing to be. Normally, part of her preparation for a year's leave was to invest in a new wardrobe consisting of the planet's current fashions. This time, she decided not to bother since the duration of her refit would be so short. Her uniforms would do just fine.

As she watched, the space elevator came alive, its glimmering lights contrasted against the increasing darkness of early evening. The dome became more visible as random points of light coalesced into a pale blue glow against the surrounding primordial atmosphere. The atmosphere was typical of the life-bearing planets found thus far: nitrogen, water vapor, ammonia, carbon dioxide, methane, and trace amounts of oxygen from limited photosynthetic activity. It wasn't Eden, but it was still a starkly beautiful planet.

As the drum spun to simulate gravity, panels in the floor offered a series of views from interplanetary crafts moving slowly past the colony or tethered to space buoys. At a certain point in the rotation Elke could see the *Lucky Strike*, now near the station with shuttle tugs already at work. A crew was removing the inbound processing modules and installing outbound colonial rings as she sipped from her glass and made herself more comfortable on her stool. The bar's subdued lighting made the bright views through the panels all the more compelling.

She sipped her whiskey as she mused about the growth of the colony. The population recently surpassed 80,000. Most of the colonists would have moved down to the dome by now. Well, it wasn't *just* a dome anymore. It was the hub of a growing complex of pressurized structures that provided living, working, and agricultural space.

Those needing to return to the drum booked a seat on a landing craft if they could afford it. The drum served as the colony's spaceport in a geosynchronous orbit above the equator, supporting interplanetary traffic throughout the local solar system and interstellar traffic like the *Lucky Strike*. The other option for traveling between the drum and the dome's base station was the space elevator, colloquially known as "the stalk." The elevator tethered the core of the rotating drum to a base station at the heart of the dome, and provided relatively inexpensive freight transportation. People could ascend or descend the stalk on their own as well, but the trip involved several days battling motion sickness and gradual loss of gravity in an uncomfortable gondola.

Elke sighed, briefly imagining walking under an open, blue sky once they arrived on Eden. She, like billions of other humans across known space, had lived all of her days in pressurized structures, on orbiting platforms, or aboard deep space vessels. She had never seen Earth, and never seen a blue sky.

Elke turned and examined the shelves behind the busy bartenders. The last time she had been to this colony, some thirty local years earlier, the selection had been considerably more modest. It had included only a handful of varieties, all locally fermented, distilled, or brewed from source materials produced by genetically engineered algae. As the colony had grown over the last three decades, it was now clearly importing some real liquor.

As Elke idly wondered which of the bottles in front of her were locally produced and whether the colony had its own breweries or distilleries, John entered the bar. He was also still in his uniform, apparently making the same judgment call as Elke. He scanned the room, saw her, and started walking her way. She caught his eye as he approached.

"Permission to join the captain?" he asked, bowing with a flourish.

Elke laughed and gently backhanded his shoulder. "Shut up, sit down, and buy me a drink!"

"Aye, Cap'n!" he said, raising a finger to get a bartender's attention. He ordered another whiskey for her and a cabernet for himself.

"Well, John, what do you think?"

"I think I'm going to regret ordering the cabernet. It's going to be disappointing. Probably made from some local, genetically engineered algal output tragically intended to substitute for grapes. There's no way they'll have vineyards at this point. I don't think there are any vineyards in any of the outer colonies yet."

"That's *not* what I meant."

He laughed, "I know, I'm sorry. Really, I think all this is amazing. You know, I started my schooling pursuing xenobiology and really loved it. But they were projecting a shortage of medics at the time, and I qualified for a pretty compelling P.I.P. scholarship, so I migrated to the Deep Space Academy at the next system, completely lost track of my family, and never regretted it. Even so, I'm still interested in xenobiology; I've read everything I can find on the life-bearing worlds we've found. I was beginning to think we wouldn't find a complex ecology. There are just so many questions. What's different about Eden? What exactly happened there that didn't happen on any of the other worlds we've found with simple organics and simple life? There's a lifetime of questions to answer. Hell, one hundred lifetimes of questions!"

He noticed Elke staring at him and realized that he was getting louder and gesturing more emphatically. He chuckled, clearly embarrassed, and broke the exaggerate persona. "I told you," he checked his volume at an emphatic whisper. "I'm still interested in the subject. This is going to be amazing!"

Their drinks arrived, and Elke lifted her shot glass in a toast.

"To Eden!"

John touched his glass to hers, sipped his cabernet, grimaced, and sighed.

Sri, Moira, and Nikko entered the bar and walked towards them. John raised a finger towards the bartender once more and gestured toward his crewmates. "A bourbon and a scotch for these gentlemen, and a chardonnay for the lady."

"Thank you, John," Sri smiled at him. "You have a gift. Do you always know everyone's preferences?"

"Only my friends…" John paused for effect, then added, "and you guys."

"Another few years with you all?" Elke sighed and rolled her eyes.

The bartender brought their drinks. Nikko raised his glass, and as the others joined him, he toasted again, "to a few more years together!"

"Might be more than a few," John said, raising his eyebrows. "It's a new colony run and I'm planning to stay on with the Eden colony." He turned to Elke and addressed her directly. "Rumor has it you've been asked to serve as governor."

It was common for at least some senior staff members of an initial colony run to form the first government of the new colony. Traditionally, the ship's Executive Officer made the return journey as Captain, along with promising junior officers now stepping into senior posts for the first time.

"Yes," Elke nodded, "and I've accepted. It'll be an interim post, as they'll be building this colony at a faster-than-usual pace. I'm hoping it goes well enough to become permanent. It's about time I settled down, and I can't think of a more exciting place. I'm glad you're considering signing on. We'll see who else does."

Sri, still standing, leaned casually against the bar. "Not an option for me, unfortunately. Somebody has to captain the return journey."

"Yes, that'll fall to you," Elke nodded again, "but you'll have some time to explore Eden before heading back."

"I'll be joining the Eden colony, at least until the colonists get settled in. That usually takes a few years," Nikko said.

"Have you been part of a new colony start before?" Moira asked Nikko.

"No, this will be my first," he answered. "I've always worked with later-wave colonists and migrants, handing them off to whomever manages operations at our destination. It's been a long time since I stayed in one place. Maybe it's time to settle down for a while."

A brief wave of nostalgia washed over Elke as she considered Nikko's response. "It has its appeal. Thinking about starting a family?"

Before Nikko could answer, the bar's relaxed mood shifted slightly as the governor of the colony entered with his entourage. Elke stood and smiled at him. The Governor caught her eye and pointed to one of the lounge areas along the opposite wall.

She nodded her understandingly, then turned back to her crew. "Sri and Nikko, our hosts have arrived. John and Moira, you're welcome to join us. You'll definitely get some free drinks, and possibly a free dinner out of it."

The group walked around one of the floor view panels and made their way toward the lounge, a warmly lit room with large windows separating it from the rest of the bar to offer a modicum of privacy. A large table with rounded corners, made of another simulated wood, sat in the center. Observation panels on the wall opposite the entrance offered a spectacular view of the planet and dome below. The remaining walls were adorned with art depicting construction of the drum, elevator, and surface settlement. The wall and ceiling texture were currently set to dampen sound reflection, which would facilitate communication between the occupants.

Governor Kelty and his companions were all similarly attired in colorful jumpsuits and jackets with elaborate patterns, which appeared to be the formal wear currently in fashion. The governor stood at the head of a table at the far end of the room.

"Governor Kelty, it's a pleasure to meet you in person." Elke raised a hand and gestured to each of her officers as she introduced them. "Please meet Sridharan Rizzo, my Executive Officer, Nikko Harris, Operations Chief, Moira Tam, Chief Engineer, and John Schmitt, Chief Medical Officer."

"Please, call me Abram," the governor said warmly, "it's a pleasure to meet you all. Please let me introduce Jenna Faulk, Colonial Lead for this expedition, George Sied, Chief Xenobiologist, David Fessler, Chief Planetologist, Cecilia Mueller, Chief Mining Engineer, and Jian Rodriguez, expedition journalist. I suggest we take our seats and get acquainted."

Elke chose the chair opposite Abram. As everyone took their seats, a virtual reproduction of the bartender that had been waiting on the *Lucky Strike* officers appeared on the table and took their orders. After the image dissolved, the group's conversation resumed.

"Why don't we go around and have everyone introduce themselves? I'll start. I've been governor here for nearly ten years. You probably knew my predecessor. She would have been governor when you left. In fact, our colony of Alexandria is now named in her honor. I was an inbound captain when she was lost in a depressurization accident. Tragic." He looked down at the table for a moment, an impromptu moment of silence for his fallen predecessor, then addressed the officer to his left. "Jenna?"

Jenna Faulk was young, but possessed a commanding air. Her blue eyes and blonde hair stood out against her olive complexion; her round face was punctuated with a prominent nose and superior expression.

"Hi everyone," Jenna addressed the group, "I've grown up here, arriving with my parents among the original colonists. I took over the family business, the largest retail operation here in the colony. Importing, exporting…You've actually been enjoying some of the adult beverages we've imported over the last few years. I'm very happy to join this team, and I couldn't be more excited about Eden."

"It's a pleasure to meet you, Jenna," Elke said. "Once we're underway, I would appreciate it if you would join my senior staff meetings on behalf of the colonists."

Jenna nodded, "of course. I'm looking forward to it."

David, a slender, fair young man with long blond hair, bright blue eyes, and a wry sense of humor, cleared his throat. "David Fessler, Chief Planetologist. I'm a pipper. I migrated here and earned my degrees from the local university, doing my field and subsequent professional work throughout the system. There's more to do here, and part of me is sad to leave, but I can't pass on this opportunity."

Cecilia, a slender young woman with dark hair, brown eyes, and full lips, spoke up from the seat next to David. "I'm Cecilia Mueller, Chief Mining Engineer. I'm also a pipper. I've been accompanying David on his field work and setting up the mining operations to extract what he's found. I agree with David that we still have work to do here—you can never really exhaust the potential of a planet like this one, let alone a whole planetary system—but this is too good to pass up."

"I'm Jian Rodriguez," said the young woman next to Cecilia. She was young and precisely dressed, slightly overweight, with dark hair and green eyes that were striking against her deep brown skin. Her face reflected pure joy as she introduced herself. "I was born here and recently completed my degree. This will be my first deep space journey. I couldn't be more excited to be chronicling this expedition as the official journalist!"

George, a scientist in his late nineties with fair skin, green eyes, and a strong build, ran a hand through his graying, light brown hair and shifted his slightly above-average girth as he prepared to introduce himself. "George Sied. Chief Xenobiologist. Ecologist, actually. I'm a double pipper! I did my undergraduate work at one of the older, more established colonies, and went on to manage their zoological and botanical parks. Never thought I'd get a chance to see an alien ecology, let alone work with one. I had just accepted a migration opportunity to come here for graduate study when we started getting reports about Eden. Right place, right time, amazing opportunity." He grinned proudly at his own good fortune.

Zoological and botanical parks were prized assets throughout colonized space. Earth's population peaked in the twenty-first century at nearly ten billion, before that century's third and most lethal pandemic reduced the population to nine billion. That pandemic, compounded by climate change, rising sea levels, famine, drought, mass extinctions, and ecological collapse, finally convinced the world that if humans didn't reduce Earth's population to a more sustainable number, the Earth would do it for them, and it would do so quite unpleasantly.

Faced with the dire alternative of mass extinction, leading global powers formed the Deep Space Cooperative, committed to spreading humanity throughout the solar system and beyond. After more than three centuries, aggressive emigration efforts, and the implementation of crushing family planning constraints, Earth's population was reduced to five billion. About two billion people made their homes under domes on the Moon, Mars, and throughout the asteroid belt. 150 million more were strewn across fifty neighboring star systems. The colonies in those systems grew rapidly through emigration from Earth and other colonies, as well as incentivized procreation limited only by human fecundity.

With nearly a third of the human population living in places that had never known a blue sky or green leaves, every colony invested in zoological and botanical parks as soon as they were large enough. The parks complemented each colony's practical agricultural portfolio with growing, diversifying microcosms of a home world they had never, and would never, see for themselves. After spending most of his life studying these developments, George couldn't wait to do his own research.

Their bartender, the human one this time, appeared in the doorway of the lounge bearing a tray of drinks and appetizers. He distributed them as the conversation continued uninterrupted.

"Elke Lubandi," she said, formally introducing herself to the group, "Captain of the *Lucky Strike*, and future interim governor of Eden."

"Captain, please don't be modest," Governor Kelty responded, "you're one of the most experienced and accomplished officers we have. This will be, what, your twentieth tour? Most captains retire after ten or so."

She tilted her head, "please, call me Elke. This will be my twenty-fourth journey, twenty-third as captain. What can I say, I love what I do. But there's nothing as exciting as leading an initial colony run."

"And you've been a governor before?"

"Yes, I spent twenty-some years as governor for one of the second wave colonies. I oversaw the initial growth, helped build the space elevator and dome, and raised a family. But, you know how it goes. You launch them as self-sufficient adults, they lead their own lives, and all too soon, it's time to think about what's next. For me, that was accepting an opportunity to captain a colonial mission."

"That's the common pattern," Governor Kelty mused, "I have a few more colonial projects that I'm passionate about, and then I'll likely start thinking about what's next, too. I envy you, Elke. I can't think of a more exciting post than Governor of Eden."

"Well, what I've been offered is an interim post," she said, "I'm hoping to make it permanent, but I imagine there will be quite a few competing for the position." She paused for a few seconds to collect her thoughts about the specifics of the mission, then continued. "It's going to be challenging. My understanding is that we'll see an accelerated cadence of inbound colonial vessels from neighboring systems. That means we'll need to make some critical decisions about how to support that growth, and we'll need to make them quickly. Do we build the usual space elevator and surface settlements? Can we build surface settlements given Eden's quakes and volcanoes? Can we, for the first time anywhere, live without a dome?"

"And don't forget," Kelty added, "those inbound ships will already be on their way when you arrive. Yes, you'll have options to evaluate, but the one option you will *not* have is to slow down or divert inbound colonists!"

After a pause, Sri raised a finger to get the attention of the table and continue with introductions. "My name is Sridharan, but everyone calls me Sri. I'm the Executive Officer of the *Lucky Strike*, so I'll be managing most of the day-to-day activities and captaining the ship when she returns so that Captain Lubandi can become Governor Lubandi. This will be my fourteenth deep space run."

John took the cue from Sri and cleared his throat. "John Schmitt, physician. I'll be running Sick Bay with a medical staff." He smiled wryly. "You'll be doing your best to avoid me, of course. I'll also be staying on Eden."

"Nikko Harris, Operations. We perform all the day to day tasks required to keep everything running. I'll be your primary point of contact for anything you need during the journey, and once we arrive I'll be staying with the Eden colony."

"Nikko," Jenna said, "I'd appreciate it if you would join my senior staff meetings once we're underway." Nikko nodded at the Colonial Lead.

"Moira Tam, engineering," Moira said from next to Nikko, "I'll be overseeing the refit and keeping everything running. And I'll also be staying."

Governor Kelty gestured to the *Lucky Strike* crew members. "Well, Captain, it sounds like most of your officers are signing on and then staying with the colony!"

Elke nodded, "yes, all of my officers and most of my crew have agreed to go. And yes, many of my current crew will be staying with the colony once we arrive. Current complement is fifty, and we'll need one hundred for an initial colony run. Forty-seven opted to go; only three preferred to stay. So, we'll need fifty-three recruits from the Deep Space Academy here in your colony, mostly operations, plus a few more engineers, doctors, nurses, and some additional security personnel. I hope that won't be a challenge."

"Not at all. The challenge will be dealing with the disappointment among the candidates who aren't chosen to be among the fifty-three. But, there will be more ships coming through to bring you additional colonists, so they'll have future opportunities. We're expecting another vessel next year."

"Excellent. We'll need to make sure everyone understands their odds of staying on at the new colony. Outbound complement with colonists will be one hundred. Since there won't be migrants to care for on the way back, seventy-five will stay at the colony to help me govern until the next colonial ship arrives, and twenty-four will return with Sri and whatever cargo the drones have managed to harvest from other bodies in the system. With outbound colonial runs, our challenge is usually finding enough recruits that want to stay. I have a feeling that won't be a problem this time."

Elke leaned back in her chair, catching the eye of each crew member in turn as she wrapped up the short briefing, "we're grateful for the extra two weeks added onto our abridged R&R time. Junior officers and crew have already started their leave. This group here will meet tomorrow morning to review candidates and fill our roster and start our leave after that. Then we will report back four weeks later for orientation and dry runs on the refit and upgrades. Moira's opted to skip the R&R and work with the harbormaster and the refit crew."

Moira nodded enthusiastically, "wouldn't miss it!"

"May I join you for that?" Jian asked, politely raising a finger once again. "I'd love to start our coverage with the refit."

"Of course. I'll be at the hangar deck at 07:00 the day after tomorrow. I'd suggest comfortable attire."

Jian nodded, beaming with the excitement of getting up close and personal with the equipment.

"Jenna," Elke continued, "as I said, my officers and crew will report in four weeks for orientation. That process takes a week, during which colonists will transfer onto the *Lucky Strike*. Nikko will be working with you to coordinate all of that."

Jenna nodded.

Governor Kelty took a satisfied look around the table. "Excellent. Well, I think we're through what limited business we needed to accomplish. I propose we continue getting acquainted over dinner."

▪LUCKY STRIKE

The flow of air was uncomfortably cool, especially where her skin still bore their sweat, but the sheets were bunched under his legs, and it was simpler to move closer to the warmth of his body.

"It's getting late; perhaps it's time to go."

"I know." A pause. "I respect that you care, but for the record, I don't. I'd prefer to stay."

"I know, but we have to keep up appearances. There comes an hour past which we cease appearing committed to our work and begin to raise eyebrows."

"I don't think that's as much of a concern as you seem to. People have needs; people spend nights together."

"Yes, but when two people, two officers, especially when there's a reporting relationship, start spending all their nights together, well, then it's more than 'needs'. Nobody cares when a couple of deep space officers, even senior officers, spend an occasional night together, but, regularly? That gets noticed, and not in a good way."

He looked into her eyes and asked, "it's more than needs, isn't it?"

She turned and kissed him. "Much more. And when we reach Eden, and we're colonial government rather than ship's officers, we can be less discreet. But not yet."

"Are you thinking about starting another family?"

She laughed, "love, that time is long past. We may be living longer than people used to, but women are still born with the same number of eggs. Mine ran out a long time ago."

"We're not old yet."

"No, but we're not young anymore either."

She paused, then sighed, looking up at the ceiling. "We're not young anymore. One day, you realize that life's pleasures no longer seem to quite outweigh the burdens and you wonder whether it's worth waking up in the morning."

After a moment, she turned again, smiling at him. "But Eden will tip the balance."

- - -

Jian, wearing athletic shoes and a colorful jumpsuit, sans jacket, leaned a bit forward and let her wrist computer guide her through the unfamiliar corridors. She didn't know her way around the top ring of the orbiting platform. She'd lived her life on the surface and didn't make it to the drum very often. When she did visit the drum, she spent her time in the lower rings.

Given that the drum was spinning to provide a sense of gravity, only the top and bottom rings have an exposed side suitable for hangar deck doors. The bottom ring, with windows facing the planet below, held senior government offices, the colonial bar, and the finest restaurants. The middle rings held the local Deep Space Academy and the bureaucracy of the space port. That left the top ring, facing away from the planet, as the only one that could hold the hangar decks for landing craft and shuttle tugs.

Her wrist computer led her to the right hangar deck. She entered the portal, walked through an airlock, and found Moira Tam dwarfed by a shuttle tug and surrounded by bulky cargo lashed to suspensor pallets.

Jian had never seen a shuttle tug before. She'd just had her first landing craft ride, which was one of the perks of her new assignment as expedition journalist, and vastly preferable to ascending the stalk. Her piece about the landing craft had been popular with her growing audience on social media. She presumed her next piece, a careful explanation of the refit, would be popular, too.

The shuttle tug's boxy gray shape loomed behind Moira. It seemed similar to the landing craft in many ways, including the fact that they each had two levels. The cockpit on the front upper deck was surrounded by view panels, and the lower deck had a cavernous space adaptable for cargo or additional seating. But while the landing craft had sleek, streamlined wings and powerful thrusters, the shuttle tug was clearly designed for orbital duties. It had a set of folded robotic arms and was festooned with attachment hooks and loops, seemingly randomly placed antennae, and a variety of small maneuvering thrusters aimed in every direction. Next to its official alphanumeric label the nickname "Betty" was colorfully emblazoned.

Moira looked up and greeted Jian as she approached. "Good morning. You're right on time!"

"You are kind," Jian blushed a little. "I'm actually a few minutes late. I had trouble finding my way."

"No worries," Moira shrugged, "we're not quite ready yet. Plus, things are still informal right now. Once we're under way, you'll find Captain Lubandi runs a tight ship." She directed her gaze toward the shuttle tug. "The hangar deck crew is almost done configuring the ship for this load. It was still configured for passengers until yesterday when they brought the last migrants and remaining crew over. There will still be seating on the upper deck for us and a few engineers, but they're pulling the seating from the lower deck to accommodate cargo. Given the aggressive refit schedule, all the shuttle tugs are engaged." She looked back at Jian, who was doing her best to keep up with the briefing. "Have you done much zero-G? There's no gravity between here and the *Lucky Strike*."

"Only between the elevator and the colonial drum," Jian said with a nod, "not something I enjoy."

"The *Lucky Strike* drum isn't spinning at the moment," Moira assured Jian. "The refit team has been working through the previous shifts carrying heavier items over and maneuvering them into place. They still have mass, of course, but they're easier to get into position without gravity. I'm not sure when they'll start spinning the drum again, or whether they'll spin it to full-G right away. You're going to find that extended zero-G and low-G time doesn't treat the fancy hairdo very well. There's a reason the officers and crew opt for these shortcuts."

Jian pursed her lips; she'd just invested in an elaborate fountain of curls and was rather proud of her ability to recreate the stylist's work more efficiently every morning. On the other hand, already thinking of herself as a deep space adventurer, she began picturing herself with a short cut, and found herself liking the idea.

Moira interrupted her reverie. "Well, while we're waiting for everyone else, would you like to walk over to where we can see the *Lucky Strike*?"

"Yes, please!" Jian agreed enthusiastically.

She followed Moira through the airlock to a waiting area with an observation panel in the floor. As they reached the panel, the *Lucky Strike* came into view. Shuttle tugs surrounded the ship, maneuvering large cargo items and moving ring segments.

They observed the work as Moira shared her wealth of knowledge about the *Lucky Strike* with Jian, who recorded everything for her publication. Jian had done her homework, watched countless videos about deep space vessels, but those couldn't do justice to the real thing.

Moira told Jian that the *Lucky Strike* was a standard configuration: a long shaft with a forward funnel to collect interstellar matter as fuel and to protect the drum. An aft funnel produced thrust as the collected matter was mixed with, and annihilated by, the antimatter reserves stored in magnetic bottles near the aft funnel. The drum, directly in the middle, consisted of standard-sized ring segments just like the ones comprising the colonial drum. Each ring, 500 meters in diameter and twenty meters thick, consisted of twelve segments grouped into four quarters.

The number of rings depended on the mission. Inbound, the *Lucky Strike* carried a complement of fifty officers and crew with 200 P.I.P. migrants, requiring six rings and leaving space for processing pods carrying mineralogical and organochemical cargo. Outbound, for a new colony run, with one hundred officers and crew, 1,000 migrants and their livestock, the *Lucky Strike* would need twenty rings for command, science labs, residential, agriculture, commercial, and industrial space.

"When we arrive on Eden," Moira explained, "those rings will be transferred to the orbital frame built by the autonomous probes already there, in anticipation of our arrival. Those probes have already been harvesting mineral wealth from Eden for quite some time. The probes' full freight pods will be loaded up on the *Lucky Strike* for the return journey where the rings once were."

Moira continued her commentary without pausing for comments or questions from Jian. She described how the drum would spin just like the colonial drum until the crew began their journey. Jian, of course, was quite familiar with how orbiting platform drums create a sense of 'down' through centrifugal force. Any time she walked through a long hallway, she could see the floor rising in front of her due to the curvature of the drum.

But Jian had never made an interstellar crossing. She wasn't used to the idea that, in preparation for acceleration, they'd stop spinning the drum and reconfigure everything so that "down" was "aft." Jian was fascinated, and knew her subscribers would be too. She knew that most of them were, like her, space natives. They never thought about gravity; they simply took it for granted when they were standing on a planet's surface or walking the interior of a rotating orbital drum. Moira explained it well:

"Imagine that you're in a motorized vehicle. When it accelerates, you feel pressed back into your seat. When you're stopping, decelerating, you feel yourself lurching forward. Now, imagine you want to reach relativistic speeds. You want to approach the speed of light in order to make the crossing in a reasonable amount of time. In this case, almost fifteen years. So, you accelerate, *relentlessly*. The whole time, just like when you're pressed back into your seat. But, because you're accelerating aggressively, constantly, and relentlessly, and because you're in space with no other gravity pulling on you, you aren't just pressed into your seat; you feel as if 'down' is aft. You do this for the entire first half of the journey, so you reconfigure the ship so 'down' really is aft. Then, when you're halfway there, you need to start decelerating. But you're not just lurching forward; you're decelerating aggressively, consistently, and relentlessly. 'Down' is now towards the front of the ship. So, before you decelerate, you reconfigure the ship so that 'down' is forward.

Of course, you can never reach the speed of light. You can keep accelerating, but something interesting happens as you approach the speed of light. Time slows down. So, even though it takes the ship almost fifteen years to make the crossing, it only 'feels' like a bit under three years for the people aboard the ship. We accelerate and decelerate at 1.2g as a compromise. We could accelerate and decelerate at 1g, the same gravity humans evolved with on Earth. We could, but the journey would take longer in real time and feel longer in the ship's time. Or, we could accelerate and decelerate more aggressively, say, 1.5g, and the journey would be a little shorter. But 1.5g is too hard on our bodies; everything, including our bodies, would feel fifty percent heavier. At 1.2g, everything's only 1.2 times heavier. We can get used to that."

Moira held up her hands. "It's a big, fascinating topic. I've...simplified a few things, but you get the gist. And we haven't touched on how we deal with higher g-forces on a temporary basis for flights between planets, moons, and asteroids within a star system, or how we factor gravitational forces into selecting which planets to colonize, or, well, a lot of other topics that could keep us here all night!"

Jian laughed, "alright, we'll leave it at that. But I reserve the right to come back with questions when I'm preparing my report to my subscribers. And, depending on how this piece goes over, I may want more!"

"Fair enough," Moira nodded. "As to the reconfiguration process itself, the entire ship is designed modularly so the reconfigurations will go smoothly, but they still take hours. Everyone and everything on board has to endure zero-G while it's going on…And if you think you don't like zero-G, wait until you see the chickens and goats!"

Jian laughed at the mental image of unhappily floating livestock. Then, with her index finger hitting the observation panel as she pointed, she asked, "why are some of the ring segments open to space?"

"Those are being purged of vermin," Moira said. She explained that as humanity spread through the stars, so did their undesirable companions. This was especially true of colonial runs involving pets and livestock. Purging the ship of vermin like rats and bugs after every journey helped, but there were always just enough air pockets for them to survive and remain a problem. Particularly troublesome were the crickets and roaches that liked to chew on communication lines, power circuits, and the ingredients of certain sealants. Every time they replaced materials with stuff bugs didn't like, the bugs evolved to find the new materials tasty. This led to communication and power outages, the occasional air leakage, and a lot of very dead bugs.

Jian and Moira's conversation was interrupted when they heard their shuttle tug pilot's voice announcing that *Betty* was ready. As they approached, Moira told Jian that the *Lucky Strike* would be equipped with twelve hangar decks and a complement of landing crafts, interplanetary crafts, and shuttle tugs. The other vehicles would be stored in compartments along the primary shaft until their arrival. Moira laughed, "Eden's vehicle needs are…special. We'll use hovercrafts instead of the usual wheeled vehicles to avoid crushing living things. The fact that the planet has an ocean means we'll also need to bring boats. We'll even be carrying a few submarines!"

Moira and Jian boarded the shuttle tug for Jian's first crossing from one orbiting platform to another. At Moira's request, a couple of the engineers moved to offer Jian a window seat. From this vantage point, she watched the pilot activate the control interface. A small virtual shuttle tug appeared before the pilot. When the pilot used her finger to raise the cargo door on the virtual model, Jian heard the real one closing beneath them.

She could hear the pilot and hangar deck operator as they prepared for departure.

"Shuttle tug pressurized."

"Hangar deck preparing for depressurization. All personnel leave the hangar deck immediately."

"Hangar deck is clear and sealed."

"Hangar deck depressurizing."

"Hangar deck doors opening, safe travels."

"Acknowledged."

The pilot held the virtual shuttle tug in her hands and gently moved it forward as Jian felt the vehicle move through the doors. Jian felt a gradual yet rapid weightlessness as the tug left the hangar deck, but it wasn't as unpleasant as riding a gondola up the stalk. The shoulder harness kept her in her seat as the shuttle tug arced forward and down. She stared out the viewing panels at the pure blackness of space, the sparkling stars, the looming planet surface below, and finally, the *Lucky Strike*, enthralling as sunlight glinted off the ship's rings. The powerful robotic arms of shuttle tugs moved massive objects as humans in bright yellow environmental suits performed the finer work required.

Finally, the pilot matched the rotation of the *Lucky Strike* drum. As they entered the hangar deck, Jian felt the pull of gravity once again. Moira leaned over to help Jian release her restraints.

Jian stood slowly and smiled. "Where to first?"

■ ■ ■

Jian and Moira spent the interim four weeks together, Jian doing her best to stay out of the way while chronicling Moira's supervision of the refit. The *Lucky Strike* was finally ready for senior staff orientation and to receive colonists. Moira, Jian, and the harbormaster waited in the control room of one of the *Lucky Strike* hangar decks at 08:00, listening to the operator talk to the incoming shuttle tug carrying the captain and the rest of senior staff.

"Preparing to depressurize hangar deck," the operator announced to the group as his hands moved across the controls. It was clear from his vantage point that no one remained in the hangar deck area, but, per protocol, he waited a full minute before proceeding.

"Depressurizing done. Opening hangar deck doors."

The doors parted, revealing the stars beyond. Moira gazed out at the colonial drum in the distance. A shuttle tug backed into the bay as she watched. The operator's hands hovered over his buttons and levers, ready to take manual control should it become necessary. It didn't.

The shuttle tug's landing gear unfolded as it settled smoothly and noiselessly into place.

"We have landed," they heard a familiar voice announce, "ready to open the cargo bay door." After a few seconds of consideration, Moira placed the voice as Captain Lubandi's. The operator closed the hangar deck doors.

"Pressurizing. Please wait." After a pause he nodded, "pressurized. You're good to go. Welcome aboard, Captain."

The shuttle tug's door opened, folding downward to serve as a bridge to the bay's working area. Captain Lubandi emerged with her senior officers and Jenna Faulk. Their group was closely followed by David Fessler, Cecilia Mueller, and George Sied. They were dressed in jumpsuits similar to the uniforms worn by the deep space officers, but without rank insignia. Jian was amused to see they'd all adopted the close-cropped hair she'd embraced during the refit.

Moira, Jian, and the harbormaster left the control room and walked through the hangar deck airlock.

"Nice flying Captain," Moira said when the groups were within earshot.

Elke chuckled, "I'd planned to do some flying while we were here, but unfortunately didn't have that chance. The shuttle operator graciously agreed to trust me with her baby." Captain Lubandi shifted into more formal body language as she changed the subject, signaling that the time for small talk was over. "Well, Chief Engineer, are you ready for us?"

Moira nodded, "my team, the junior officers, and the crew are all waiting for us on the bridge. The refit is complete, and the harbormaster is ready to hand the vessel back to you. We'll spend the week getting everyone up to speed." She turned to address the colonial leaders, "welcome aboard."

"Thank you," Jenna said with a smile, "we're looking forward to bringing our people and cargo over."

"The ship is yours, Captain. Safe travels. I wish I could join you," the harbormaster said to Elke.

Elke turned and addressed him graciously. "Thank you. You'd be welcome." They shook hands, and the harbormaster boarded the shuttle tug.

"Let's get started," Elke said, gesturing to the airlock door. "Nikko, I'll leave Jenna, Cecilia, David, and George in your capable hands. I'm sure they'll want to inspect their quarters and facilities."

"Moira, are you ready for me to escort them to the colonial rings?" Nikko asked.

"Certainly," Moira said, "you'll find that everything is shipshape and in the usual places."

Nikko turned to the colonial leaders, "please, follow me." The leaders bid farewell to the captain and senior crew and left the hangar deck.

Elke turned back to Moira and asked, "are we walking, or are the lifts working?"

The drum was equipped with six lifts evenly spaced around the drum rings. They moved up and down when the *Lucky Strike* accelerated or decelerated; but now, while the drum was rotating, they moved laterally fore and aft.

Moira laughed. "They're working. Everything is working. Shall we?"

Moira and Elke walked to the nearest lift, Jian trailing behind at a respectful distance. They passed technicians working at open panels with parts and tools at their feet. The senior officers followed.

"You're sure everything's working?" Elke asked again.

"Yes, Cap. We're still buttoning up a few loose ends, but nothing to be concerned about. Overall, the upgrades are enhancements and refinements. Nothing transformational. Some enhancements are impressive; engine efficiencies are improved almost 6%. It reflects some impressive advancements in matter/antimatter reaction theory; we had to replace almost a third of the drive system! Anyway, Amar, your navigation interface has been upgraded to take this into account."

The group reached the lift. Elke examined it and sighed. "You weren't tempted to brighten things up while we were gone?"

"Sorry Cap," Moira laughed a little, "regulations haven't changed. But, as always, you're welcome to decorate your quarters as you see fit. You'll like your new comms station, by the way. Bigger interface, better sound and color. We've also upgraded personal food fabricators, so you'll find you have more selections with richer seasoning options. Everything else in our quarters is pretty much the same."

"Anything in Sick Bay?" John asked.

"Again, nothing transformational," Moira turned to face him, "the diagnostic panels are upgraded. The displays are bigger and brighter, and they display more data. They're supposed to read the patient faster. The holographic patient projection is more detailed, and you have more interactivity options. You have a variety of new nanobot-managed microsurgery capabilities, including greater precision and more autonomous functions that you can initiate. More routine procedures can now be performed through the projection interface. They're all good, and I think you'll appreciate them, but nothing transformational."

John nodded, then joked, "no cure for the common cold yet?"

Moira shook her head. "No, but maybe they'll figure that out while we're underway." She laughed lightly at the aside, then continued with her briefing:

"We'll be distributing the latest wrist computers. Huge improvements have been made in their processing capacity and projection quality, of course. They're thinner and more comfortable. You'll see that they're physically smaller, but they actually project a broader holographic surface. There's better integration with ship's services and wearable devices now, too. I've found that the latest assistive interfaces are more responsive and possess more comprehensive monitoring of what you're doing, so they're better at predicting what you might request. If you're like me, you'll be frustrated the first day or so as it learns you, but you'll love the convenience of the device once that part's over."

As the lift station doors parted and the group stepped into the lift car, Moira switched back to the broader subject of move-in logistics. "Nikko's team is taking care of moving everyone's gear from the shuttle to their quarters. By the way, Sri, they found your memorabilia and it's already waiting for you in your quarters. It turned up with some of the refit materials. Not sure how that happened."

Sri thanked her and breathed an audible sigh of relief. He loved Earth history and had an extensive collection of old Earth seafaring artifacts, including his sextant coffee mug, a G.P.S. receiver, and various nautical flags and charts. He'd been heartbroken when it seemed his collection had been lost in transit.

"Anything else notable?" Elke asked.

Moira shrugged. "Like I said, assistive interfaces for wrist computers have been improved. They're also upgraded at our bridge workstations, but I haven't seen much of a difference there so far. Maybe it'll become more evident over time. Unfortunately, like the wrist computers, they'll need to relearn you, but they can do that during the dry runs. Subspace communications have been improved, so we can now send and receive at higher velocities. The responsiveness is better, too. Still not quite an interactive conversation, but better than before.

There have been a number of maintenance upgrades. Much better augmented reality. When we look at a panel, we can pull up various levels of detail about what's behind the panel and, from there, pull up supporting diagnostics, documentation, and so forth. Every time we go through a refit, there's more embedded intelligence in our subsystems to interact with augmented reality features for richer diagnostics and improved prognostics."

The lift stopped and the doors parted. They walked through corridors that seemed very familiar to the crew members, yet gleamed like new.

"They did come up with a fundamentally better way to modularize everything to reconfigure 'down'. It's hard to describe, but I'd be happy to demonstrate when we're preparing to reconfigure for departure. It's always been a challenge to reconfigure all twenty rings in one shift for an outbound colonial run, so this will make a huge difference. It's pretty clever, actually." Moira gave a sheepish grin as she admitted, "now that I've seen it, I'm embarrassed it never occurred to me."

They reached the bridge portal and stepped through into a large room with a low, round table in the center. A complementary structure above it held a variety of projection devices. A series of unique workstations were positioned around the table, each equipped with the interfaces that individual officers would use most often. There was plenty of space for the officers to push their chairs away when they preferred to stand. The workstations were separated from their neighbors by gaps that would allow the officers to step up and interact with projections. Surrounding the table and workstations was a raised platform with additional supporting workstations and display screens. The ship's complement, except for Nikko and the members of his team engaged with preparing for the rest of the colonists, stood on the platform. Every audible conversation came to an abrupt halt as the captain and senior officers entered the bridge.

Jian entered tentatively; Moira pointed at her, then nodded towards the raised platform. As Jian found her way, she glanced back and was heartened to see Elke smile and nod.

As Elke and her crew found their stations, a detailed holographic projection formed across the whole table. The image showed the surface of the planet below, complete with dome, space elevator, and colonial drum. Shuttle tugs moved across the *Lucky Strike*'s rotating drum. Three interplanetary crafts, one heading toward the colonial drum and two away from it, flew across their field of view. As they watched, one of the crafts crossed the edge of the image and disappeared.

Jian cleared her throat. Elke turned to her and asked, "any questions from our journalist?"

"Could someone explain the display?" Jian was relieved at the opportunity for clarification.

"Certainly," Moira answered without missing a beat. "As you can see, we're looking at the colonial drum and everything around the spaceport, all of which is necessary to safely navigate until we reach a safe distance for initiating our interstellar journey." Moira shifted her attention to Elke as she continued, "the local Deep Space Academy was kind enough to compile a series of dry run scenarios for us. I haven't seen them yet. One of their engineering students loaded them, Captain, and we can start the scenarios whenever you're ready."

She turned back to Jian and added, "don't be concerned about what you're about to see. The dry run scenarios will involve a number of things that aren't actually happening: near misses with other spacecraft, that sort of thing."

Elke thanked Moira, straightened her posture, and clasped her hands behind her back as she looked around the room at faces familiar and new.

"Good morning," she said, "I am Captain Elke Lubandi. I know some of you well, and I'm honored by your willingness to serve with me again. I'll get to know the rest of you as we work together. I'm sure you've done your homework and know my reputation, just as I have reviewed your files, so for the moment, we're acquainted. I am impressed with what I've read, and honored that you have chosen to take this journey under my command."

She paused, looking at each face in the crowd in turn, "I love new colony runs. There's nothing like escorting the very first colonists to a new star. Nothing like nurturing a new home, and, indeed, a new hope for humanity. And, there's never been a colonial run like this one. Eden is unprecedented. This new planet holds such promise. It will finally allow us to not only finally ask our questions about truly alien life, but have those questions answered.

Well, people, it's time to go to your stations. We'll be conducting dry runs and going through scenarios. If there are any problems from the refit, we'll discover and address them. After a week, our colonists will come aboard and get settled in. At least, as settled in as 1,000 colonists can be. Then, we'll depart."

She looked through the crowd and made eye contact with one of Nikko's senior team members. "Mitch, please take Nikko's station while he's working with our colonial leadership. John, we'll start while you're heading to your Sick Bay."

Then she stood at attention, once again addressing the crew as a whole. "Dismissed," she projected to the crowd.

Junior officers and crew saluted and filed out of the bridge to go to their stations. Elke turned to her small group of officers still standing nearby. "Well, people, it's time. I couldn't be more thrilled by this opportunity or more delighted to have the pleasure of your company on another adventure." She looked up and said, "load the first scenario."

PART 2:
ARRIVAL

•19 DAYS BEFORE ARRIVAL

Jian woke up slowly in her modest quarters. She had one, relatively large room containing a comfortable bunk, a kitchenette, a closet, and a generous work surface with comms screen. A small door to the side housed a private restroom with a sink, toilet, and shower. Her room was located in the colonial residential ring closest to the two command rings, providing her proximity to facilitate the interviews she'd been conducting since they'd begun their journey to Eden more than two years ago.

She sat up slowly. She'd spent the previous evening a few levels aft in a colonial speakeasy indulging in a touch (or three) of moonshine. There'd been a small, pleasant crowd celebrating approaching Eden, including a particularly attractive young farmer from the agricultural rings. She smiled, remembering how she'd enjoyed flirting with him, then frowned slightly because she wasn't quite sure why the interaction hadn't progressed beyond flirtation. She wasn't worried about it; she figured their paths would cross again, given the initial size of the colony. Besides, she'd recently started enjoying the amorous attentions of a certain young geochemist. She thought about him tentatively, then realized he'd have started his shift by now. Perhaps she'd suggest dinner. *Yes*, she thought, *that would be lovely.*

The *Lucky Strike* was still about three weeks from Eden, but today was a special day for the crew. Sometime today they would be decelerating enough to resume subspace communications for the first time in fourteen Earth years. Like most of the people on the ship, Jian was looking forward to reconnecting with those she'd left behind, more than a little anxious to see the difference a decade and a half had made for them. She'd spent a fair bit of time exploring living with relativistic travel with the more experienced officers and crew during her interviews, realizing that would be popular with her audience. It fascinated her to consider the degree to which everyone's lives would have changed. Given the number of personal connections she had left behind and the number of years that had elapsed, there was a high probability that several of her friends and family would have experienced significant life events while she had been on the ship. They may have become parents, achieved a degree, or even passed away. She shook her head as she shifted her weight to the edge of the bed and placed her feet on the floor, organizing her thoughts about what was on her agenda for the day. She needed to finish curating her interviews to prepare them for subspace transmission to her audiences once the *Lucky Strike* decelerated enough.

She stood up, stretched, then pulled on a robe before padding towards the restroom.

Sri had been her best interviewee on the subject of relativistic travel. Given his fascination with old Earth sea faring, he'd put it all in perspective. They'd met in his spacious quarters early in their crossing, once they'd accelerated beyond any ability to exchange messages. As Executive Officer, his responsibilities often included one-on-one meetings, so he had a small conference table next to the display case holding his memorabilia. He had leaned back in his chair, fingers knitted behind his head, and shared his very interesting perspective on space colonization as his eyes skimmed over his collection of seafaring equipment.

"In many ways," Sri pondered aloud as Jian took notes, "humanity's spread through adjacent star systems is not unlike the way the Lapita people spread through Oceania some 4,000 years ago. They settled one island at a time, spreading through nearby islands until they eventually populated thousands of them, large and small, throughout the Central and Southern Pacific." He shifted his eyes towards hers, shivering with excitement before continuing: "our crossings in deep space vessels are a lot safer and, of course, more comfortable than their ocean crossings in outrigger canoes."

Jian recalled prompting him to expand on the differences between space and ocean travel.

"Well, it's worth noting that, centuries ago, crossing the Earth's oceans with wind-powered vessels was a perilous adventure. Voyages of two to three years were not unusual. With adulthood starting in one's teens and many adults geriatric by their forties, such a journey was a substantial fraction of one's adult span. In modern times, we live a lot longer. What's our average life span these days? 150? 170? I don't know how much of that is Darwinian selection in the aftermath of the twenty-first century and how much of that is medical advancement, but the result is ultimately the same. A modern crossing of ten or fifteen years is comparable to the two to three year journeys of our ancestors. Plus," he'd added with a wink, "the magic of relativity making such a crossing feel like two or three years doesn't hurt."

Sri talked about the deep space community and the kinds of people that were drawn to the lifestyle. It was a financially rewarding life, with substantial pay that earned interest over the long years they spent in flight. Many did a handful of crossings, retired to start a family and live a more conventional life. A healthy share of those found themselves drawn back to deep space after their children left the nest. Then there were the lifers: people who limited their relationships to only the bonds between shipmates for the duration of each journey, never needing more lasting connections. Jian thought the latter seemed lonely, but could not deny the magic of a life spent witnessing profound changes over centuries in year-long slices. She thought of the changes Captain Lubandi must have witnessed when she was coming of age in the Belt over 300 years ago. She had seen the beginning of interstellar flight and led many new colony starts throughout various waves of humanity's expansion.

Her series on relativistic travel would need to be ready to transmit soon, and she couldn't wait to share these insights with her readers. Of course, the most anxiously awaited transmissions would be about Eden and the rest of the surrounding solar system. Jian was as excited for these updates as the rest of the residents of the *Lucky Strike*, but she wouldn't have much information about those right away. The scientists aboard were still working with the data sent by the deep space probes, and, since those probes predated subspace communications, the data set available to them at departure had been fourteen years old. They had been analyzing that data during their journey, and Jian had learned a lot from their analyses and compiled some interesting interviews, but she also knew that the researchers at Alexandria continued to receive data while they traveled. By now, they had an additional fourteen years of data that the crew wouldn't receive until they decelerated enough for subspace communications. Jian looked forward to interviewing the scientists over the next few weeks as they blended the work they'd done in transit, the fourteen years' worth of missed data, and the fresh data they would collect directly from the probes themselves once the ship reached Eden.

As she made her way to the kitchenette to prepare her morning tea, Jian shifted her thoughts to the fact that there wasn't much to report about their actual travel. She'd been intellectually prepared for sustained periods without obligations, but she had not been emotionally prepared. She had spent most of her time on the ship compiling profiles of the colonists. The pieces were human interest stories that some of her readers would undoubtedly enjoy. They detailed various colonists' opinions on careers such as agriculture and crafts-manship, skills the colonists were learning and interests they pursued during the flight, and how some were indulging in projects they'd long procrastinated on and finally had no excuse not to complete. She also wrote articles about family life in transit, including courtships, births, and conflicts. Each piece was more compelling than the last, and she thought her readers would enjoy reading these accounts of ship life as they made their way to a new colony start.

She'd interviewed Amar Aggrawal early in the journey, and his inter-esting personality still stuck out in her mind. He was easily distracted with a tendency to go off on tangents, but incredibly intense regarding anything to do with interstellar travel.

"There really isn't much that needs doing once we leave the solar system," Amar had casually explained during their first interview. "We've been making interstellar crossings for more than three centuries, and nothing worth men-tioning ever happens in deep space. We're really only busy at three points of flight: as we're leaving, as we're approaching, and as we're reconfiguring from acceleration to deceleration."

The ship certainly had been buzzing with activity during the reconfigu-ration as they prepared for launch. Moira had been right about the chickens, goats, and pigs not liking zero-G! It had been especially interesting to see how the aquaculture team sealed the fish farm tanks until gravity was restored. The tilapia handled it better than the catfish, but none were overjoyed at the shift. But, once they were underway, they quickly achieved 1.2g and the ani-mals relaxed.

"What's it like to cross deep space?" Jian had asked Amar during another interview. "What will we see when we approach the speed of light?"

"Initially, there's some doppler effects. Ever hear something rush by you? The pitch is higher as it approaches because the sound waves are compressed as the source approaches, then, as it's moving away, the pitch gets lower? Well, there's a similar visual effect as we accelerate or decelerate. Colors appear to shift as our speed affects the perceived frequency of light coming from nearby stars. But, that's an interim phenomenon. As we get closer to the speed of light, you really can't see much of anything. It's quite disorienting, really, and we close the observation ports for most of the trip. There are some very good simulations you can access from shipboard systems if you'd like."

Jian sat at her table, absentmindedly working her way through her break-fast. She'd had some interesting interviews with Daniella about deep space security. Essentially, her team kept the peace among the colonists. There were no conflicts between colonies, at least not beyond the home system, where tensions between Earth, the Moon, Mars, and the Belt had long since reduced to a light simmer. As extrasolar colonies grew, so did their probability of developing a criminal element. Colonies established long enough for regular shipping between domes, orbiting platforms, and outposts were sometimes plagued by interplanetary piracy; but, at least thus far daunting logistics pre-vented interstellar conflicts. And there were no external threats, no signs of alien intelligence, although the discovery of Eden suggested that one day they might find a world with not only a complex ecosystem but also intelligent life.

Otherwise, Moira and her team made sure systems kept running, John and his team kept everyone healthy, and Nikko and his team did whatever it took to keep the colonists happy and productive as they conducted their business, lived their lives, and generally prepared for their new home.

Jian pushed away her dishes and sank back into her chair. She needed to think through what to publish *after* her series on relativistic travel, until she had fresh Eden material. She'd come up with something.

Her comms panel reminded her it was time to prepare for Captain Lubandi's Monday morning meeting, and she didn't want to be late. She'd stopped attending the meetings a few weeks after departure; the meetings got progressively shorter and the discussions progressively more routine as the ship settled into the journey. She'd start attending again this morning now that they were approaching Eden and the discussions were sure to be more interesting.

She loaded her breakfast dishes into the ultrasonic washer and began preparing for her day.

● ● ●

Amar Aggrawal strode purposefully, en route to the captain's conference room for the Monday Morning Meeting, weaving his way through junior officers and enlisted crew as they made their way with just a little less urgency than his own. It was a shift change for them; for him, he was late. Again.

He shook his head; it was frustrating and a bit embarrassing. He had been running some simulations through his wrist computer, preparing for the meeting, and lost track of time. Again. He checked his wrist computer to monitor progress. The simulations should be completed soon.

He was getting close. In fact, just ahead, he could see Captain Lubandi as she ducked to enter the conference room portal. At least he wasn't that far behind her this time.

Amar reached the translucent portal and entered the spacious conference room to the sounds of senior staff making small talk. Jenna Faulk was asking, "so, we'll be receiving a message from the drones? How will that work?"

Captain Lubandi took her place at the head of the triangular table. The senior officers present today included Executive Officer Sridharan Rizzo, Chief Engineer Moira Tam, Security Chief Daniella Wu, Operations Chief Nikko Harris, and Chief Medical Officer John Schmitt. Also present was Colonial Civilian Leader Jenna Faulk; she'd been attending their meetings since departure.

Now that they were approaching Eden, Elke had asked Jenna to invite the Chief Xenobiologist George Sied, Chief Mining Engineer Cecilia Mueller, Chief Planetologist David Fessler, and Journalist Jian Harris. All were present as well.

After returning greetings, Elke turned to Moira and asked, "well, Moira, do you want to take a stab at answering Jenna's question?"

Moira leaned forward on the table. "We began sending deep space probes long before we were capable of interstellar travel. The probes deploy drones to survey systems and send back the data we used to prioritize targets for colonization. With each wave of colonization, we sent more probes further and further into neighboring systems. The probes that discovered Eden were sent over a century ago, so they predate subspace communications. That's why it took so long to receive the first survey results."

Next to Jenna, Cecilia nodded her head in thanks. "I've been wondering about that. How did they know where to aim their signals?"

"Excellent question," Moira nodded enthusiastically. "We're talking about fourteen light-year's away; they can't simply 'broadcast'. As we sent probes, we programmed them with the locations of existing and planned colonies, enabling them to target their messaging. They actually send their messages to a number of destinations."

David said, "okay, that makes sense…but why haven't we been receiving signals since departure? We have the data sets we'd accumulated before we left, but nothing new since."

"That's a question for Amar," Moira redirected. "Amar?" There was no response.

"Amar?" Some of the senior crew exchanged telling looks.

"Amar!"

He blinked and looked up from his wrist computer. "Oh, excuse me. Right…What was the question?" David repeated himself as Elke cleared her throat in obvious frustration. "Well," Amar answered, "I wouldn't blame you for thinking the optimal flight path to Eden would be along the radio transmission, so you'd think we'd be able to continue picking up the signals. But that turns out not to be the case. You can't pick up conventional radio signals while traveling at relativistic speeds. Even if you could, well, two things. First, our journey is long enough that the relative movements of solar systems have to be taken into account. Second, since we're approaching but can never reach the speed of light, we have to take those relative movements into account when plotting our optimal course and when aiming radio transmissions, so our course is not fully aligned with the transmissions. If you'd like, I can prepare some simulations to help."

David said, "I would be curious, but another time. So, how is it that we'll have a subspace link today?"

"Before we left, once I'd plotted our optimal course, we launched a relay probe. It's been flying twenty-three days ahead of us and reached the probes exploring Eden's system a few days ago." A light went off on Amar's wrist computer, cutting him off abruptly. He turned his full attention to it.

After an awkward pause, Moira cleared her throat. "Right…so the relay probe has arrived by now, established conventional communications with the deep space probes, and knows where we'll be as we approach. As we speak, it's been downloading all the data the probes have been collecting from their survey drones, and as soon as we've decelerated enough, will begin sending us that data. As we receive it, we'll make it available to you and your teams."

"Why does it need to know our locations?" David asked.

"Subspace doesn't work like a radio message," Moira replied. "Basically, the way subspace communications works is that the message is transferred instantaneously from source to a destination, then travels the way transmissions normally travel. So, the relay probe that's been traveling ahead of us needs to know where we'll be at any given time in order to transfer the message to a place where we'll be able to receive it as a conventional message. That's also why we have to decelerate enough to be able to pick up the message; there's a relationship between our velocity and the precision of the placement. Now, that's…a really gross simplification, but I think it works, doesn't it?"

Elke said, "that ability to 'transfer' the message instantaneously is also what gives me hope that, someday, we'll finally get past the speed of light." She laughed, "of course, Moira keeps telling me that I'm dreaming, but can't really explain why."

Moira laughed as well. "You don't have the math. Essentially, it's one thing to transfer an energy sequence; it's quite another to transfer matter. Now, you're right, in theory; I expect, one day, we'll find a way to travel faster than light. But nothing in the literature suggests we're anywhere close to a plausible concept let alone a prototype."

Elke nodded and clapped her hands. "Well, be that as it may, it doesn't affect anything we'll be doing today or in the near future. Amar, what's our status?"

Amar, having lost track of the conversation again, looked up from his wrist computer and blinked slowly. "Oh, right. Well, we're doing…well." He paused and flicked his finger across his wrist computer; as he did so, a projection of the *Lucky Strike* appeared over the table, with the star system they were approaching appearing in front of Captain Lubandi. "We are…" He paused to consult his wrist computer. "…on track. We've been traveling for 764 days ship's time, which is, actually, sixteen days less than we would have before the refit engine upgrades, which is, really, amazing. Real time, it's been fourteen years and sixty-eight days. Anyway, we're still decelerating and… nineteen days from arrival." He looked up, then tapped his wrist computer and the projection dissolved.

Elke paused for follow-up questions. When no one volunteered any, she thanked Amar before addressing the group as a whole. "Okay, we have nineteen days to prepare for arrival. Any pressing status updates before we deal with that? Sri?"

Sri waved and said, "nothing pressing; all's under control."

"Colonists and crew have been remarkably healthy this past week," John commented. "A colonist tripped during a squash match and broke his wrist, but that's about it."

"We had to break up a minor scuffle between a couple of colonists that had a little too much moonshine," Dani added. "All in all, we've had very few incidents, so I assume our official position regarding whomever is running the still, or stills, is to continue looking the other way, right?"

Nobody said a word, but a few people smiled. Dani's face broke into a slow grin as she said, "very good. Personally..." Dani lowered her voice slightly, "I think it's rather good moonshine."

Nikko smiled at Dani's comment. "Nothing to report from me," he said. "The colonists have been keeping busy with their day to day responsibilities; we've been able to respond to their requests in a timely manner. Jenna, do you agree?"

Jenna nodded, "absolutely. But we are all anxious to get to work."

Moira looked towards Cecilia, David, and George. "You've been doing your best with the data we received before departure, right? But I'm sure you're anxious for more?"

George sat back and looked at David and Cecilia; David nodded to him. George said, "right, we have that substantial data set we'd received before departure. We have preliminary survey data from each body in the system. Once we identified Eden's ecosystem, we limited ourselves to long distant observations of Eden itself. The rest of the system is being thoroughly explored. We have some of that data from before departure, and we're more than ready for the relay downloads."

Elke said, "well that's why I asked you to join us. Moira, when will we start receiving reports from the relay probe?"

"Later today," Moira said definitively. "Don't forget, we should also be receiving a data compilation from the colony based on what they collected via conventional transmission from the probes while we were in flight."

Elke turned to the planetary scientists. "Here's how this will work. You'll get what's forwarded to us by the colonial science team at our point of origin. As for the relay probe, by now, it will have arrived and engaged with the probes exploring the system. It will start sending us subspace messages laden with data collected since we started this journey. Get ready," she smiled, "I'm sure it will be a substantial set." David smiled and rubbed his hands together. Cecilia laughed. George grinned shyly and side-eyed the other two.

Elke continued, "we'll be receiving progress reports from the probes. They should have antimatter generators in near solar orbit generating fuel for Sri's return trip. They should have the colonial frame ready for us, orbiting above the most stable point they could find along Eden's equator with a new command ring and additional agricultural capacity. And there should be processing pods ready for Sri to ferry back with some of the preliminary automated mining of the system. All of that will be of interest to my staff and me. There should also be substantial data from the surveys of the non-Eden bodies in the system, which will be exciting for the planetological team. George, I'm afraid there will be limited data for you beyond long range observational data."

George nodded, "understood, and I wouldn't want it any other way given the circumstances."

Jian raised her hand. "You said the colonial frame will be orbiting above 'the most stable point along Eden's equator', why is that?"

"At some point," Moira responded, "we'll be building a space elevator from the colonial frame to move freight between the planet's surface and the colony. That requires the colony to be in geosynchronous orbit...ah...in an orbital path that is continuously above the same place on the planet surface. That will need to be along the equator. And it requires that the place on the planet surface to be as geologically stable as possible, not an easy thing to find on Eden. Now, we could build the framework in an orbital platform that isn't synchronized above a particular point, but it would be challenging to move it later, so we might as well build it in geosynchronous orbit right from the start. If you're really interested in the details, I'd be happy to send you some reference materials."

Jian nodded eagerly. "Please!"

After a moment, Elke continued. "The relay probe will work both ways. The probes have drones ready for closer examination of Eden, but they'll need programming. You and your team can do that from here, and the relay probe will convey your instructions."

George was nodding, "yes, we've started that work."

Elke said, "very good. I want to remain cautious. As we approach, we'll let the drones get incrementally closer but let's not attempt anything close to the planet's surface until we're in orbit."

David raised a finger. "Actually, we're hoping to deploy some seismic probes. The data we have suggests a great deal of seismic activity on Eden; we'd like to learn more about the tectonic plates, how they're moving, etc. It's critical, really."

Elke raised an eyebrow. "Why?"

David placed his hands on the table, leaned back, and paused. "Okay, *my* turn to get technical." He smirked, to Elke's annoyance. "The surveys from the other bodies in the system indicate relatively high levels of extremely rare metals with significant commercial value for various industrial uses. They're rather high on the periodic table. We're *very* excited about that.

Usually, when these metals are present at all, they're merely trace elements; but here they're present throughout the system in unusually high levels. Our current thinking is that there must have been a stellar explosion relatively near this system early in its development, seeding the system with far more heavy elements than we've seen in other systems. You can bet we'll be arguing about that for years." He smiled at his own joke.

"But can they be practically extracted? That's where tectonic processes come in. As the drones have explored the other bodies in the system, we haven't seen indications that the metals have been concentrated in many places. If they're not concentrated in ore bodies, they're much more challenging to extract. So far, we haven't seen the kinds of tectonic activity elsewhere in the system that would form readily mineable ore bodies. Well, except the asteroids, the deep space probes have been working on some of the asteroids. There are some promising deposits there."

Cecilia nodded along. "Yes, yes, it's really quite exciting! I've been modeling how to economically extract these metals from some of the other bodies. It can be done, but if tectonics have concentrated any of these metals on Eden, well, it'll be a game changer."

"Of course," George frowned, "I'm not happy about this. We need to be *very* careful about any plans to exploit Eden's mineralogical potential. However, given how dynamic Eden seems to be, gathering the seismic data will help David's team build models that can predict quakes and eruptions. We are very interested in studying the ecological impact of those events, and of course, such models have great safety value for landing parties."

Elke nodded. "Okay! Here's what we'll do. You are the colonial representatives of the Xenobiology and Planetological Guilds, but as Captain and then Governor, I have the last word on this, understood?"

Cecilia, George, and David exchanged glances, then nodded.

"Excellent. I assure you I won't make any decisions without consulting each of you. David, I completely understand how hungry our colonies are for rare metals, but I must heed George's concerns. For the time being, I assume there would be other geological riches to be had, yes?" David raised his head and gave a quiet "mhmm.".

"Okay," she continued, "George, I also completely understand that Eden is a unique and precious resource. While I'm open to safely capitalizing on all of Eden's resources, I'm also committed to preserving the only pristine, complex ecosystem in all of known space. Acceptable?"

George hesitated, then leaned towards Elke. "Yes, for now, but I'm still concerned about reaching an agreement on what constitutes 'safely.'"

"That is fair, we can discuss details later." She turned to Cecilia, "you're the mining engineer. We'll be relying on you to help us understand, fully and thoroughly, what any extraction process would entail, and what recovery processes would be used for restoration once the metals are extracted. Work for you?"

"Of course!" Cecilia responded so quickly she almost cut Elke off. "Of course."

Elke sat upright in her chair and for a rare second, her face became rather serious. "For now, nothing touches Eden. I'd like a briefing, as we start the drone surveys, on your proposals for deploying seismic probes safely. We must not contaminate Eden with any Earth-sourced organic materials. Acceptable?" Murmurs of agreement echoed from around the table, except from Amar who was still distracted by his blinking display.

Elke relaxed and leaned back. "Great. When we're in orbit, I'll be seeking input from my superiors, as I'm sure you will. Let me be clear: I am absolutely fine with you consulting with your superiors, and I'm absolutely fine with you deciding, if you disagree with any of my decisions, that you want to escalate issues to my superiors. All I ask is that we are always completely open, honest, and transparent about all of this. Understood?" Silent looks of respect filled the quiet.

"Alright, next topic." She looked from George to Cecilia to David. "Here's what I'd like. I presume you and your teams will have meetings to review what you're learning as we receive data, and I'm sure those discussions will be quite technical. I'd really appreciate it if you would also hold regular briefings, less technical briefings, for those colonists and crew that are interested. I know I'd be a regular attendee. I have a feeling that we'll have a large group."

Cecilia glanced at George and David, then smiled. "It would be a pleasure and a privilege, Captain."

"Excellent. Nikko, gauge crew interest so we have an idea how large a venue we'll need for these, what should we call them? Lay briefings? General briefings? I like that better. Jenna, would you, please, also gauge interest amongst the colonists?" Both Nikko and Jenna nodded happily. Elke paused, then turned to Moira. "I suspect we'll have quite a crowd, especially at first and when we arrive. You may need to do some reconfiguring." Moira pursed her lips and began making notes on her display.

Elke placed her hands on the table, "next, and I believe last, topic. Soon we'll be reconfiguring to rotate the drum as we prepare to move ring segments to the colonial drum framework. Moira, Nikko, I'm hoping it'll go a little more smoothly than the reconfigure from accelerating to decelerating?"

Moira grimaced, "yes, Captain." She glanced towards Nikko. "As you recall, we did a detailed review of that transition, and we've addressed the issues."

"Yes," Nikko shifted in his chair nervously, "that's right. I'll be working with Jenna to make sure the colonists are better prepared."

"Very good," Elke nodded, "I'll hold you to that. No chasing terrified livestock through the residential ring segments this time. In fact, I'd like to review your corrective action plan. Can we do that later today?"

Nikko and Moira exchanged glances. Moira said, "yes Captain, at your convenience."

Elke laughed, "you don't need to get that formal. Let's just work together to get it right this time."

She looked around the room. "Anything else? No? Okay, we're done here. Let's get to work!"

▪10 DAYS BEFORE ARRIVAL

George was in his lab, preparing for the first general briefing. It had been a very exciting week. He knew the accumulated data from the probes would have very little to offer him, but the drone surveys, once programmed, proved amazing.

He remained distressed about the planetological surveys of Eden. They'd agreed on a seismic probe strategy; it wasn't challenging, really. The deep space probes would manufacture the seismic probes from materials harvested elsewhere in the system. The manufacturing process would be conducted in space, ensuring sterile probes. But the seismic probes did support more detailed mappings and the development of some predictive models. He had to admit, the models were delightful. They were predicting a major ocean floor quake and subsequent tsunami that would impact one of the northern hemisphere continents in the not too distant future, and that would create an excellent opportunity to study ecological succession on Eden.

But the success of the seismic probes led to another worry for George. Now the planetologists were asking to drill. George understood the need to analyze drill cores to evaluate ore bodies, but they were far, far more intrusive than seismic probes. At the very least, they needed to establish some qualifying criteria to minimize drilling

He needed to think through a compelling argument against mining Eden. After the ecological devastation of the twenty-first century, after hundreds of years of restoration efforts, Earth's ecosystems were, finally, showing signs of relief. But true recovery of all that was lost would never be possible. Gene banks were helping restore some of the species lost to extinction, but they had their limits.

Someday, some long, distant day, when evolutionary processes had time to do their work, to diversify and expand the limited foundation that survived humanity's unchecked growth, emerging ecological complexities will approach what early humans experienced. But it will be a different Earth, as different as the Earth of humankind different from the Earth ruled by dinosaurs.

Man's despoiling of Earth should not be replicated on Eden; God's granting of a second chance must not be squandered.

He needed to come up with a plan. But not right now. Now, he needed to go to the briefing.

He left the lab and walked through the corridor. The briefing room had been set up in an adjacent ring, so he needed to pass through a bulkhead door to get there. As he approached the briefing room, he was pleasantly surprised to see an extensive crowd. He made his way through, entered the room, and walked to the dais in the front. There he found his team waiting with David, Cecilia, and their teams.

As he approached, Elke noticed him, and walked to the front of the room. She was a commanding presence; as she walked, the room quieted, and people took their seats. She stood at the front and looked across the room. Her voice effortlessly filled the room.

"Welcome! I am absolutely delighted to see this level of interest in the scientific part of our journey. Of course," she smiled, "I recognize there isn't much else going on right now." Light laughter trickled throughout the room.

She turned to the scientists sitting on the dais. "George, David, and Cecilia, would you introduce yourselves and your teams?"

Cecilia spoke first. "Certainly, Captain. Hello! I am Cecilia Mueller, Chief Mining Engineer. I've brought two members of my team to help me answer your questions: Scott Boucher, mining geomechanics, and Olga Busch, mineral processing." They each raised a hand and nodded as their names were stated.

David cleared his throat and stood. "I am David Fessler, Chief Planetologist. I've brought five members of my team: Sun Zaytseva a volcanologist, Enrique Ford a seismologist, Esteban Kay a geochemist, Terry Castenada a hydrologist, and Sylvia Ogilvy an oceanographer."

A hand was raised. Elke pointed in recognition. "Yes?"

"Can you, please, explain these titles?"

David nodded, "of course. I'm a geophysicist, which means I study the planet holistically, looking into the overall structure." He turned, pointing to each member of his staff as he spoke:

"Sun is a volcanologist; that's easy, she studies volcanoes. Eden has a lot of them. Enrique is a seismologist; he studies quakes, and Eden has a lot of those, too. Esteban is a geochemist; he studies the planet's chemical composition. Terry is a hydrologist studying how water moves over the surface and through the rock, and Sylvia is an oceanographer, which I think is pretty self-explanatory. That's a new specialty actually; we haven't needed an oceanographer before." He craned his neck, trying to spot whoever asked the question. "I, uh, hope that helped. Thank you. Um, George?"

George smiled, then turned towards the eager crowd. "And I am George Sied, Chief Xenobiologist. I'm an ecologist by training, so I study how living things interact with each other and their environment. I spent most of my career managing zoological and botanical parks. I brought my entire team. Hasani Owens is a wildlife biologist, so he studies how populations of living things interact in the field. Gustav Bang, physiologist, so he studies living things in terms of how they work at the tissue and organic level. Rhett Sandoval, cytologist, which means he studies living things at the cellular level. Jerry Romano, anatomist, so he also studies how living things work, but more mechanically. Sam Garcia, evolutionist, which I think is pretty straightforward; he'll be trying to figure out how the living things on Eden are related to each other. Gwen Babangida, microbiologist, so she studies tiny things. Marita Park, molecular biologist, so she studies how living things work chemically. Kamala Prescott, geneticist, so she will be studying Eden's life forms and their genetics. Martin Walters, botanist, also pretty straightforward and with whom I think our farmers have become acquainted, and Hank Patel is a behaviorist, who'll be studying the living things of Eden and understanding why they do whatever we discover that they do." He sat back down and nodded to Elke.

Elke nodded back, "thank you everyone. All right, let's get started. David, would you like to give an overview of the system?"

David stood and said, "my pleasure!"

He flicked his finger across his wrist computer and a projection of the solar system appeared above his audience. As he walked through the room and pointed at the projection, he began, "the sun is unremarkable, very similar in size and brightness to Earth's sun. In fact, at first glance, the entire system is reminiscent of our home system and many of the others we've colonized. Eight planets including Eden, third from the sun. Two gas giants, an asteroid belt, smaller planets in the outer positions. Nothing unusual there!"

He stopped walking and spread his hands. "So, what *is* different? Obviously and, as George will review with you, Eden has the only complex ecology we've found. What else? Well we know from the drone surveys that the system is unusually rich in precious metals, rare minerals, and other desirable materials. We don't know, yet, how easily these materials can be extracted. We are hopeful though. The complex tidal forces from Eden's moons are extremely likely to have produced rich ore bodies that we should be able to economically access." He laughed and added, "we're going to be embarrassingly rich, my friends!"

After a brief eruption of applause and cheers, David answered a few questions about the solar system and Eden's tectonic activity. Once the questions petered out, he tapped his wrist computer and the projection dissolved.

As David took his seat, George stood and flicked his finger across his wrist computer. A large, rotating Eden appeared at the front of the room. As he walked towards it, he took a deep breath. He knew that the prospect of great wealth from the system's mineral and organochemical resources were a powerful motivator. Profits pulled colonists outward as surely as Earth's draconian regulations pushed them towards the stars. But Eden herself was special; Eden must be cherished and preserved. The burden was on him to capture their imaginations. He needed them to understand; he needed them to *love* Eden.

"While David and Cecilia are studying the entire system, my team and I will be focusing on Eden itself. It's really a very, very beautiful planet. We've been flying drones in successively lower passes, getting more detailed looks at the various ecosystems of the planet. We have not flown any drones low enough to directly observe life forms, but we'll do that once we're in orbit and can initiate landing parties. I'm looking forward to taking some samples." George began his next comment, but stopped abruptly when he saw a hand slowly raise. "Oh, um, yes please?" He surreptitiously glanced at Elke to gauge her interest in his presentation, but she was whispering something to John.

"Do we expect animals and plants?"

"Great question. We expect things that are the *equivalents* of animals and plants. See the patches of green? Something is harvesting solar energy to produce food and oxygen. The atmosphere is thirty-one percent oxygen, and that doesn't happen without photosynthesis. We also know, or at least believe, that 'green' is the most efficient color for photosynthesis based on what we've observed on Earth.

However, on Earth, the green comes from chlorophyll. We've found primitive cells on other planets similar to Earth bacteria, and some of those cells photosynthesize, and some of them are even green, but none of them have chlorophyll. And while we've never found a complex ecosystem before, we have found life before, and life on every planet has been chemically unique. All life on Earth uses the same four nucleic acid base pairs to build DNA: guanine, cytosine, thymine, and uracil. On every other planet that we've found life, the planet's nucleic acids are unique. Will we find something analogous on Eden? Sure. But not the same." George paused, trying to gauge the crowd's interest. He hadn't lost them already, had he?

"So if you're thinking of popular culture stories with aliens falling in love with each other and raising children that are half human and half whatever, you're actually more likely to produce children with an earthly potted plant than anything you'll find on Eden." Light laughter rippled across the room. George smiled with them.

"But that's a GOOD thing! The fact that we're fundamentally incompatible chemically with Eden's organisms may allow us to walk openly on the planet's surface. For example, none of Eden's microorganisms should be able to infect us because we are biochemically incompatible. However, we intend to be *extremely* conservative, and we're going to be very careful with our first expeditions.

Okay, so, I'm going to play a few flyovers for you. Here," he quickly swiped his finger across his wrist computer, "is a drone flyover of what we believe is a 'rainforest.'"

The Eden projection was replaced with a quickly moving broad expanse of green, giving the audience the feeling that they, like the drone, were flying over a broad, pristine forest.

"Now that looks like trees with leaves, but they're absolutely *not* trees. Earth trees have trunks composed of wood, which is made of cellulose. From an ecological point of view, it makes sense that we would see living things competing with each other to gather light, so they would have physical structures allowing them to hold photosynthetic surfaces analogous to leaves high above the ground like a 'rainforest' canopy. But I guarantee that, when we inspect one, we'll find the green is not chlorophyll, and whatever the structure is that is supporting the photosynthetic surfaces, it won't be wood."

Another hand shot up. "If Eden trees *look* like Earth trees, will it be the same for the animals? How will they compare with Earth's animals?"

"We expect to see living things, 'creatures', of all sorts that can move independently through the environment. So, ecologically, they will be analogous to Earth's animals. But again, we expect them to be wholly and distinctly alien, and cannot wait to see them and figure out how they work."

He shook his head and smiled broadly. "I cannot find the words to express how excited we are about Eden. Everything will be entirely new. I'm expecting millions and millions of new species, all entirely different from Earth. For millennia, everything we managed to learn about biology has been from just one example. Now we have two examples! Now we can compare and contrast, and discover which things are Earth-specific or Eden-specific, and begin to understand which things might be common to all ecosystems."

As he spoke, the rainforest flyover had been replaced by more flyovers. Beautiful landscapes moved past them. Snowbound mountain ranges, temperate forests, rivers, deserts, oceanic expanses filled with shoals of colorful, swimming marine life. Everyone in the room was captivated, losing track of the passage of time. Some, having lived at more established colonies, had seen zoological and botanical parks. The extent, and variability, of these parks varied with the maturity of the colonies, but nothing compared to the broad, pristine stretches of natural environments flying past them as the projections continued. Nothing they'd ever seen recorded on Earth could compare, either. Certainly, one could imagine what the landscapes had been like before humanity's spread, before cities, before agriculture, before mining and logging and climate change and mass extinctions. But that could not compare with the splendor displayed before them.

As the recordings played, a hand was raised. "Do we know why Eden has a complex ecology, when no other planet has had one?"

George sighed deeply. "That is a profound question. Truly, it is *the* question! We don't know, at least not yet. Maybe we'll never be completely sure. But understanding, as much as we can, what factors favor the evolution of a complex ecosystem is essential to our search for more complex ecosystems. Perhaps, even, a complex ecosystem with intelligent life!"

He paused, then continued. "Consider how many worlds we've found in the 'sweet spot'. That right distance from the right kind of sun to have liquid water and an atmosphere. But nothing more than organic material or simple cells like Earth's bacteria. No complex cells like an amoeba. Nothing multicellular, nothing big enough to see with the naked eye. If we can determine why Earth and Eden have complex life, perhaps we can improve our ability to find more Edens. Or, perhaps, to appreciate just how rare and precious such worlds are." He walked towards his audience, sweeping across their faces, speaking with an almost religious fervor. "What is it? What happened here? Will we find Eden has complex cells with nuclei and organelles? Or something completely different? Is it sexual reproduction, which we know accelerates evolution? Does Eden life even have sexual reproduction, and, if yes, how different will it be than anything we've seen on Earth?"

He shrugged, then said, "let's go back to how green the planet is. Let's dig in, for just a moment, and talk about how incredibly exciting that is!"

He paused, raising his hands, and then said, "we know green is the most efficient color for photosynthesis. That's why, on Earth, green things, green because of their chlorophyll, won the evolutionary competition. Green things clearly won the competition here, but we know that they won't have chlorophyll. The biochemistry of every world we've explored has been different. Yes, there will be amino acids and nucleic acids and fats and so forth, but they will be unique to Eden just as they've been unique to every other world. But, even though Eden's biochemistry will be unique, evolution *still* sifted through it all until something green emerged and won the competition! This teaches us so, so much! What else? In what other ways has evolution, starting with fundamentally different raw materials, produced similar results? Or completely different results? My friends, Eden is the most exciting discovery in history! In a few, short days, we'll begin harvesting answers to questions we've never been able to ask before!"

George caught himself, and sheepishly looked around at the colonists. "Sorry…I uh, didn't mean to be long winded. It's a big, fascinating question. It's almost like, once that special something occurs, this amazing proliferation of living forms becomes not just possible but inevitable." He smiled, a look of wonder on his face, "just…fascinating."

The recordings had ceased playing while George was talking, and the lights had brightened. Elke stood, paused for a moment, and said, "George, thank you. Cecilia and David, thank you, and thank you to your teams. This has been amazing and inspirational. I, for one, feel so privileged to be part of this expedition, and I can't wait to see what we discover next."

She turned and faced the audience. "I hope you'll join me in expressing appreciation for the presentations. They are the result of a considerable amount of work analyzing large volumes of information." A large round of applause rose up from the crowd; the three scientists beamed at their evident success, especially George. Elke continued, "we are still receiving large amounts of informative data, so I propose we have another presentation in, say, a week?" She raised her eyebrows as she faced the scientists. Cecilia nodded yes for the group.

"Very good. And, after that, we'll be consumed with reaching orbit and transferring our drum rings to the colonial platform." A few groans. "Now, now, it can't be helped, and I believe it will go much more smoothly this time." She smiled, "thank you once again for coming, we'll reconvene after we reach Eden."

The crowd began to leave the room, a few people moving towards the scientists to ask additional questions. She saw John talking to Gustav and Marita. She caught George's eye and, walking over, asked "do you really think we'll be able to explore Eden without environmental suits?"

"I don't know, but it's possible. However, to be very clear, I refuse to rush this process or rush a decision. We'll be doing extensive research before anyone breathes Eden's air."

"Good. I appreciate your caution." David and Cecilia approached them, and Elke smiled. "I don't know about you, but I could use a drink. There's a speak-easy I'm not supposed to know about down the corridor. How about we go see whether we can blend in with the general crowd?"

Cecilia laughed and said, "Captain, with all due respect, you're not going to blend in if you can't enter without ducking. But I'd be happy to join you, and the first round is on me." All were smiling as the presentation hall emptied of its last bodies.

▪ DAY 4

Nikko sat at his comms station reviewing progress reports with classical music playing softly in the background. He'd loved music as long as he could remember, soothing himself during stressful periods by practicing the keyboard. He particularly loved the Baroque and Renaissance eras, and during the crossing he found a few kindred souls amongst the colonists. With a little help from Moira's printing team, they'd managed to fabricate reasonable, quality instruments and put on a few concerts. He was currently listening to a recording of one of their better performances, Handel's 'Water Music', for which he'd taught himself to play bassoon.

His operations team accounted for forty members of the crew, now Eden's colonial government. A handful of them would be part of the *Lucky Strike*'s return complement, but most would take on important roles in the developing colonial society. Working with Moira's engineers, they'd successfully transitioned the *Lucky Strike* from deceleration to drum rotation, and then managed the transition of colonial rings from the ship's drum to the colonial framework constructed by the deep space drones in anticipation of their arrival. The transition went as well as could be expected, only running an hour longer than projected. Livestock escapes were minimal, and only a few succumbed to stress. There had been a handful of complaints from colonists, mostly regarding breakage. Overall, he was quite proud of how well it went. Given the scale and scope of moving 1,000 colonists and multiple residential, agricultural, and industrial rings, the word "smoothly" would never really be applicable.

He'd hoped for a bit of a break before organizing a landing party schedule, but waves of xenobiologists and planetologists were anxious to get to the planet's surface. He and his operations team were doing their best to accommodate them. He smiled. It was like dealing with small children excited about new toys.

Part of that accommodation had been accelerating lab space construction while they were approaching Eden. It was standard practice to build labs on the hangar decks used to accommodate landing craft. There was ample space, moving samples from landing craft to lab space was convenient, potential contamination risk for the rest of the ship was minimized, and everything could be easily purged by opening the hangar deck doors. The challenge was making the xenobiology lab space substantially bigger than standard; Eden's samples might be quite a bit larger than the usual organochemical or single celled samples taken from other planets.

Nikko's wrist computer buzzed gently, reminding him it was time to start walking. The first landing party was waiting for him to conduct checkouts. They'd conducted environmental suit training during their outbound journey, but everyone was required to pass a final checkout before setting foot on the planet. As casual and informal as Nikko might be off duty, he took his operational responsibilities extremely seriously, especially with regard to life support equipment. Whether in space or on the surface, people's lives would depend on their environmental suits. Hanger deck lockers for life support packs would be properly stocked, suits would be properly maintained, and anyone leaving the safety of the pressurized orbiting platform would have to demonstrate environmental suit proficiency to his satisfaction. There would be *no* exceptions!

The first landing party consisted of the colonial leadership (rank had its privileges) and members of their direct staffs that won their respective lotteries. Governor Lubandi was waiting with John Schmitt and Daniella Wu (Nikko, unfortunately, had not won a spot for this trip), Jenna Faulk with Jian Harris, George Sied with Hasani Owens and Gwen Babangida, Cecilia Mueller with Scott Boucher and Olga Busch, and David Fessler with Esteban Kay and Sun Zaytseva. Elke was offered the honor of being the very first to step foot on the planet, but she called for another lottery instead. George Sied had won.

Nikko found Mitch and two operations team technicians waiting when he ducked through the entry portal for Hangar Deck 6. He greeted the landing party, then turned to Mitch and said, "okay, let's get them checked out!"

■ ■ ■

As the landing craft hovered centimeters above the ground, the forward hatch lowered, and George Sied walked down the ramp. He hesitated at the end, slowly extended his right foot, and became the first human to take a step on Eden. It felt *wonderful*.

Soft, gray, moist sand squelched softly beneath his boots as he walked away from the landing craft. He could hear gentle surf rolling in behind him. He looked to his left to see exquisitely blue waves breaking on the shore, foaming white as the waves crested and collapsed, turning gray as they rolled and churned the sand. As the surf receded, small, tiny mouths opened and closed in the settling sand as air bubbles surfaced and burst. Rare wisps of white clouds, stark against the spectacular blue skies, floated high above the surface of a deeper blue sea extending to a horizon accented by two of Eden's moons. Sunlight sparkled on the light waves that textured the water near the shore. Further out, the smooth surface stretched like glass to the horizon.

He'd read about such things; he'd seen videos. He'd dreamed about what it would be like to experience them for himself. None of this prepared him for the sublime beauty, the magnificence, of this reality.

Throughout his life, his experience of true wilderness was limited to botanical and zoological parks bounded by their enclosing domes. It was unnerving to look outward, to see 'wilderness' extending to this horizon.

He desperately wished he could open his helmet to feel and smell the breeze. He wondered what it would feel like to remove his boots and walk out into the surf, to feel the ocean surge about his toes, to feel his feet sink into the sand as the water undermined them. To his right, the sand rose gently to a rocky crest that obscured whatever lay beyond, smooth and unspoiled save for debris abandoned by the receding tsunami. He looked back as the rest of the landing party followed him. He saw his boot prints, and he was saddened by the first signs of human incursion on a virgin wilderness.

He heard Elke say, "we're out of the landing craft. We should be ready to return in a few hours." Through the environmental suit's intercom, the landing craft pilot responded, "very good. I'll be tracking the transponders in your suits and monitoring this signal. When you're ready for pickup, please move to an open space and let me know."

The landing party spread out as the landing craft moved gently upward, then accelerated forward and banked left over the open sea, increasing in altitude until it faded from view into an immense blue sky. The sun was a few degrees above the oceanic horizon, looking very much like depictions of Earthly mornings.

George thought that sky might be the prettiest thing he'd ever seen.

He sighed, then said, "Gwen and Hasani, let's do a little reconnaissance. We'll get more structured in our sampling later. For today, let's just take whatever we come across."

David came trudging up the beach through loose, dry sand, then crouched to inspect a rocky outcropping. He took a small hammer from his kit and broke off a sample. He stood, holding the sample in his fingertips against the sunlight. "Igneous, granite, quartz, mica, nothing exciting. Esteban, are you taking some sand samples?"

Esteban was standing in the surf, staring out at the horizon. He responded, "was planning to. George, you picked a beautiful spot."

George smiled, "thank you." He watched the waves lap at Estaban's feet, and tried not to count the growing number of boot prints in the sand. "We can recreate a number of ecosystems in our zoological and botanical gardens, but we can't recreate…this." He paused, then continued, "this is where the tsunami hit a few days ago. Picture it: the water suddenly receding as far out as you can see, exposing the seabed, then a twenty-meter wall of ocean rushing inland. The hand of God sweeping everything away up and down this coast, twenty kilometers inland."

Dani, standing on the rocky crest above the sand, let out a low whistle. "Hard to imagine now! How often did you say these events occur?"

Sun interjected, "the drones monitoring the planet reported at least two or three off-shore quakes in a typical year, resulting in significant tsunamis wiping out a shoreline somewhere on the planet. There are also inland quakes and volcanic eruptions. It's a dynamic world."

Hasani added, "we're expecting to find a significant number of organisms adapted to enduring and exploiting such disturbances."

George started walking up the incline, noting that the dryer sand was more challenging to traverse. As he approached the crest, he could see rock outcroppings wiped clean. He looked back to see numerous large and small lumps of organic matter strewn across the beach, presumably stranded by the tsunami.

"George, take a look at this." He turned and saw Gwen crouching by one of the lumps, examining it with a pair of forceps. It was a generally tubular shape, about twenty centimeters in length and about five centimeters in diameter, a silvery blue in color. "Seems to have some internal skeletal structure. I'm also guessing this is the front; this looks like a mouth but it's the strangest mouth I've ever seen. There are…flaps of tissue on the body; they remind me of fins, so I'm guessing this was something that inhabited the water column like a fish."

George crouched by her. "Looks small enough to take as a sample. Note that there's nothing trying to eat it; I'm guessing the tsunami wiped away whatever life forms normally inhabit this beach area while stranding these… things on the shore. It'll be interesting to see how quickly scavengers arrive." He stood up. "For now, we'll take advantage of the gifts the tsunami left us for some preliminary sampling of aquatic life forms. Once we dissect a few things, we'll get more of a feel for Eden's life forms and build a strategy. Look for plant equivalents."

George stood and looked out across the sea. He slowly scanned the horizon, imagining, once again, the tsunami building and washing over the beach. He heard Hasani say, "I don't see anything plant-like; I think it was all washed away."

Cecilia walked inland, climbing a gentle incline of exposed rock. She reached the top and yelled back, "hey! You need to climb up here and look at this!"

George followed, and as he climbed, noted debris scattered as far as he could see. The slope increased in the distance and he could see dense foliage starting at what he presumed had been the high-water mark of the tsunami. "That looks like it might have been a plant equivalent's trunk."

Elke laughed over the intercom. "George, if we all promise to not forget this isn't Earth, can we just say 'plants' and 'animals'?"

George chortled, "sure, why not?" He walked over to the trunk and knelt, taking some tools out of his pack. It was too big to bring back intact today, but he could certainly take samples from it.

▪DAY 5

The world was green, and damp.

Lighter and darker shades of green. Relentless green. The leaves were green, the debris was green, albeit decaying to yellow and brown. The trunks were the only relief: a dark brown nearly indistinguishable from the shadows. There were no bright colors, and nothing like earthly flowers, at least not that they had found yet. In a world of relentless green, the animal life was also green: green to evade predators, green to surprise insufficiently wary prey.

A drone flew a few meters above the team, scanning in every direction.

The ground was soft and uneven, covered with fallen, decaying material that obscured trippable hazards; the wet soil squelched under their steps, sucking at their boots as they continued their walk. In places, the ground was wet enough for their feet to sink deeply. The forest sloped gently upward and downward as the xenobiologists trekked through the trees. Hasani Owens, the resident wildlife biologist, employed his tracking skills as they navigated around deep gullies and large knolls. The canopy was thick; little light penetrated to their level. It was a small blessing. The gloom constrained undergrowth that might, otherwise, impede their progress and interfere with their line of sight.

As they walked, tiny things jumped out of their way. Little creatures rustled amongst the rotting debris and small critters suddenly darted away as they approached. They could hear movement in the canopy above them. Occasionally, something scurried up or around a tree trunk. There was a continuous buzzing sound from small, insect-like things flying all around them.

In their early cataloging of Eden's living things, they found that, like Earth, there were highly diverse groups of creatures without internal skeletons. Many lacked skeletons at all: things without limbs, things with bristles, things with tentacles, things with fluid shapes. These things burrowed through the decaying debris, grazing, scavenging, and hunting.

But, like Earth's insects, spiders, and their kin, the most prevalent creatures had exoskeletons. Once evolutionary experimentation stumbled upon this body design, here on Eden and back on Earth, there seemed no limit to their diversity and numbers. But there are limits to how big things with external skeletons can grow, so they were complemented by a plethora of creatures with internal skeletons. They slithered through the leaf litter, scrambled up and around tree trunks, and flitted through the canopy far above them.

So far, the scientists hadn't encountered anything large, but just to be sure, their patrol drone kept flying above them, circling around them, monitoring the area, searching for anything big enough to be a threat. Another, heavier drone flew behind them, just a meter or so above the ground. The rain forest was too dense for a hovercraft, so this drone was carrying their sample containers. Periodically, as they walked, one of them would see something interesting. Their 'little helpers' improved their ability to recognize life forms and compare them to their growing catalog. The little helpers were adapted, by Moira's team, from systems used by planetologists conducting surveys. They highlighted life forms by outlining them on the xenobiologists' visors as they were detected and recognized, then provided, via heads-up displays, relevant information about what the user was seeing. New discoveries led to rest breaks as samples were taken and placed in one of the containers.

A distant, crashing sound came from their right, and they froze.

Their patrol drone darted over. Hasani lifted his hand and they waited for the drone's report. So far, they'd not encountered large predators, but that could certainly change now that they were exploring more ecosystems. Their environmental suits were impenetrable, but they were flexible to support freedom of movement, and therefore unable to protect them from potential physical damage in the event of an attack.

The drone reported that something large was approaching them from the east. Hasani spoke softly, "let's each stand next to a tree, then activate your camouflage." The scientists separated, their bright yellow environmental suits shifting to match their background. Once they were all situated and nearly invisible, they turned their attention towards the eastern gloom.

In the dim light, they could see movement before detail. As it slowly emerged, they could see that it was large, some five meters in length, a bit bigger than an elephant. The dark body was massive and roughly cylindrical, with rough skin and a short, conical beak at the front and, to the rear, two thick, stubby tails were held with a light upward curve. Its six, short, stout limbs held the body well above the ground as it walked. Each limb had two articulation points reminiscent of knees and elbows and another, lower articulation that allowed the feet to swivel. The six digits of each foot were arranged in opposing sets of three, either grasping structures or simply splaying across the forest floor. From the back, directly behind the beak, was another, dorsal limb ending in a sensory cluster with what appeared to be ear flaps and six 'fingers' ending in eyes.

As it slowly progressed, two of the eyes were aimed forward, over the beak, and the others moved independently of each other as they scanned in every direction. Occasionally, one of the eyes would stop moving, and other eyes would shift to scan in the same direction. Sometimes, when this happened, the creature would pause as the dorsal limb shifted to align the sensory cluster with whatever caught the attention of the scanning eye.

It stopped by a thinner trunk, and its beak opened into three sections as it began taking bites from the trunk. They could hear the fibers of the trunk snap and sever as it bit into and worried the trunk. After achieving a notch, it reared up and began pummeling the trunk with its forelegs. With a sharp crack, the trunk broke near the bite, and the relatively small tree fell with a crash, its canopy settling down within a few meters of the landing party. The beast walked towards them and nuzzled the canopy. Its sensory cluster continued to scan as it trudged over the fallen tree. The beak opened again, and multiple tongues extended and grasped at branches. The animal pressed itself forward into the tree. As it did so, a muscular mass extruded through the mouth, past the grasping tongues, and pressed into the leaves. It held this position, moving gently, adjusting very slowly, making small grunts and snuffling sounds. The tree creaked and strained as the tongues pulled on the branches.

Hasani was the first to dare speak. "What are we observing?"

Hank Patel, behaviorist, answered in a hushed tone. "Clearly a feeding behavior, but full of surprises. Some of the smaller creatures we've collected and examined had beaks and tongues, so I was expecting it to use the beak to snap off branches and the tongues to pull them in. But it looks like it's everting its stomach and pressing it against the leaves to digest them!"

They walked slowly and carefully to the left, improving their view of the animal and tree. They could see the leaves and smaller branches dissolving against the moist, muscular tissue, material flowing slowly and gently towards an opening at the center of the mass as the muscle rippled and pulsed. They watched the tongues stretch to grasp branches and pull the tree against the muscular mass.

Jerry Romano, anatomist, commented in wonder, "that's clearly digestion! It's everted its digestive organ and is digesting and absorbing the leaves."

Hasani asked, "wouldn't it lose a lot of moisture feeding that way?"

Sam Garcia, evolutionist, replied, "well, sure. I would expect a lot of moisture loss from the stomach lining, but we're in a rainforest. I'm sure it has no trouble replacing the water. But I doubt we'll see similar feeding strategies in desert or even temperate ecosystems."

Hasani stared wide-eyed. "Fantastic! This drone footage will be fantastic!"

As they watched, Jerry said, "did you notice that the beak has three segments? The smaller specimens we've examined suggest some of the life forms have trilateral symmetry rather than bilateral symmetry!"

Hank responded, "you mean, instead of a right and left like we have, they have a right, left, and top?"

Jerry nodded, "exactly. I mean, this is all preliminary, we've only begun examining samples we've collected, but it looks like some of the body designs are, as you say, right, left, and top. Absolutely fascinating! Eventually, we'll study embryonic development to learn more. Well, I presume we'll study embryonic development; we haven't figured out, yet, how anything on this planet reproduces!"

The animal paused, and the stomach and tongues retracted. The beak snapped at some of the larger branches, now bare, that were impeding access to leaf tissue. When these branches were disposed of, the tongues extended, grasped branches, pulled the tree closer, and the stomach everted again. It took the animal a few more passes to fully consume the tree's photosynthetic surfaces. It released the rest of the tree, retracted its stomach and tongues, closed its beak, and after scanning the area with its sensory cluster, crouched on its rear legs as dark brown and white material fell to the ground beneath its lifted tail. Then it continued walking through the rain forest.

As it moved away, the group deactivated their camouflage, and Hasani exclaimed, "I think we've earned our pay for the week!" He turned to Martin and said, "take some samples from the tree it was feeding on. And take some samples from the droppings."

Martin crouched and picked up a leaf. "These leaves are amazing. Have you looked at these leaves?"

Hasani hesitated. "…they're leaves, Martin. *Leaves…*"

Martin sighed. "Typical wildlife biologist! Plants are just there to nurture your precious animals! But my world is full of wonders, too, my friends! Here, look." He held the leaf up and activated his little helper, then shared, to each of their heads-up displays, a slightly magnified view of the leaf.

"See? It's a leaf, sure. It's green and broad and thin like any Earth leaf. But it just doesn't look quite right, does it? It's not just that it's different from any species you've ever seen before. The vein pattern is weird. The petiole is weird. It's a little too thick, too, and without the waxy surface you're used to on thicker leaves on Earth succulents."

Martin's little helper zoomed in as Martin continued. "Now, look closer. Remember your undergraduate botany classes? Who can tell me what's missing?"

He paused. Then, as nobody spoke, he shouted, "TIME'S UP! No stomates! The little holes Earth leaves have? The ones that facilitate transpiration? In a rainforest species, no less! Completely different! Completely alien! Without stomates…I have *no* idea how these things draw up water unless they have some alternative to transpiration…"

He paused, then, chagrined, apologized. "Sorry! It's just completely different. And that's this super-kingdom; the trees from the other super-kingdoms are completely different from this! Anyway, to your question, as you noticed, it walked by a number of similarly sized trees. I can't help but wonder what was special about this one. I'll need to check against our other botanical samples."

Hasani responded, "exactly."

▪DAY 6

Governor Lubandi ducked to enter her conference room. It was in the first command ring constructed by the deep space probes. The room's overall configuration was very similar to the room she'd left behind on the *Lucky Strike* for Captain Sri, who was currently sitting at her table representing *his* ship. But this conference room would never be reconfigured for acceleration, creating the opportunity for observation windows facing the planet surface.

They were positioned above the most tectonically stable point along the equator to facilitate the eventual construction of a space elevator. Unfortunately, that stable point was in the middle of a desert at the center of Eden's largest continents. Still, from her seat, Elke could see not only the desert below, but also where it gradually transitioned to savannahs and then rain forests in the distance.

It was morning, and she could still see the dawn line retreating to the west as Eden turned eastward towards the sun. There were clouds to the north; rains were falling on the forests but did not reach the desert below them. *Beautiful...*

Also present were the colonial leadership and her colonial government department heads transitioned from their *Lucky Strike* roles. Everyone was still clad in their deep space uniforms, though they were no longer wearing rank insignia. She imagined that, over time, there'd be a gradual migration to current fashion, whatever that might be. She found herself anticipating Amar's belated entry, then remembered he would be staying aboard the ship for her return journey. She'd miss him. She chuckled out loud at the thought.

As she took her seat, she called the meeting to order. "Welcome to our first colonial staff meeting. Let's start with engineering. Moira?"

Moira leaned forward and said, "everything's in perfect working order, obviously, or we couldn't have conducted the transition of the colonial rings.

We've also completed checkouts for all the daughter craft from the *Lucky Strike*. Landing craft and interplanetary craft are either in their colonial hangar decks or tethered to nearby buoys. Two interplanetary craft are currently deployed with planetologists and mining engineers inspecting automated mining operations in the asteroid belt. Landing craft are conducting regular trips to the surface for xenobiology and planetological survey teams."

Nikko raised a finger, "right. The two interplanetary craft left as soon as we finished the deceleration reconfigure, per David and Cecilia's requests. We're allowing planetologists and mining engineers with proper certifications to pilot interplanetary craft, but my people are piloting landing craft and surface vehicles.

We are now taking xenobiology and planetological survey teams to the planet surface daily. We're making sure everyone gets checked out on environmental suits before boarding a landing craft. Surface vehicles have not yet been deployed; I expect to make them available over the next few days. It's been a bit of a scramble."

Elke nodded, then turned to George, Cecilia, and David. "And I'm confident that the leaders of the xenobiology and planetological teams will be buying you drinks as soon as the colonists get a pub up and running."

David and Cecilia laughed heartily. George smiled, "first round will be on me; we are grateful for the accommodation. We're also delighted with the 'little helper' capability. We've been initially focused on the site of the tsunami predicted by the tectonic models." He glanced around the room. "You may recall that was the site we chose for the first landing party. The tsunami happened as we were approaching, and our landing parties were there to observe succession starting with the initial disturbance. In fact, we'd like to explore setting up an observation blind."

Elke asked, "succession?"

"Ecological succession," George responded, "is when a significant disturbance occurs, a tsunami, fire, quake, whatever, and fundamentally alters an environment, that environment undergoes a series of transitional phases as it returns to its original, mature state. On Earth, you might see a wildfire clear a forest. From the ash, you'll see a grassy meadow, followed by shrubbery and young trees, and, ultimately, a return to the mature forest. Eden's tectonic activity results in regular disturbances, so we feel it's essential to understand succession on Eden. Events like this happen regularly from a planetary perspective, but we can't be sure when we'll have another chance to observe a succession on this scale from the moment of the initial disturbance."

She nodded and asked, "and an observation blind?"

George answered, "we would like to establish structures with life support at the initial landing site that would allow a rotation of ecologists to continuously monitor the ecological succession."

Elke paused, her face becoming serious with thought. "I'm not ready to build structures on Eden. But I fully support getting landing parties to the site on a regular cadence." She turned to Nikko and added, "I very much appreciate that you and your team have been so accommodating, but we need a more orderly approach."

Nikko nodded, then said, "agreed. I'd like to have the xenobiology and planetology team's landing party plans locked and loaded for at least a week in advance at all times, so I'll need your planning people to talk to my planning people as soon as possible. We'll be ready for that this afternoon."

Elke placed her hands on the table. "Next?"

"Nothing notable from security," Dani yawned. "Everyone's been too busy settling in to get into trouble."

John commented, "a few minor injuries associated with transitioning ring segments from the ship to the colony. Nothing major." He turned to Moira and Nikko and added, "having gone through one or two of these before, my compliments."

As they nodded, Elke added, "yes, well done. In fact, the smoothest I've seen."

Jenna commented, "it was my first, so I have nothing to compare it to, but yes, I can see it went very smoothly. Thank you Nikko. The colonists are settling in. We lost some livestock in the process, so the butcher shop currently has a surplus of fresh meat. We expect livestock populations to recover quickly. We harvested all the crops before we finished decelerating. Now that we don't have to plan on re-orienting 'down' again, we can establish more extensive agricultural space combining what we brought with the additional space the probes had waiting for us." She smiled and added, "I don't know about you, but I'm looking forward to switching from algal-based synthetics to the real thing for the crops that don't grow well hydroponically. Also, with Moira's and Nikko's help, all of our industrial facilities are up and running again, and the shopkeepers are looking forward to no more disruptions until we need to expand the colony."

David spoke next. "We very much appreciate the early checkouts for our teams. The two interplanetary craft completed their inspections of the automated mining operations in the asteroid belt and are evaluating additional promising sites for the next round of mining operations. Cecilia's people verified the processing pods were ready, and they are being towed by drones back here to be added to the *Lucky Strike* frame for her return journey. As to Eden, while we were approaching, we conducted drone flights and identified a number of promising sites for landing party evaluations. We've been collecting samples that are now being analyzed; we're hoping to identify candidates for drill core sampling."

Elke nodded and said, "Very good. And I'd like to see and review your proposals for drill core sampling. Let's be *very* clear that we are reviewing proposals at this point, and that George and his team will be part of the review process." She leaned back, "when…could we have another general briefing? Our colonists have been asking; they're quite curious about their new home!"

George and David looked at each other, and David responded, "we can be ready in a couple of days. George?" George began to nod, then corrected himself, "actually, we'd have more to talk about in about a week. How about then?"

"Excellent." Elke turned to Sri and smiled, "Captain?"

Sri smiled back. "Thank you, Governor. The *Lucky Strike* is on track. I would like to thank you all for your recommendations regarding the returning crew complement; I'm running simulations with my new senior officers and staff. We'll be ready soon but won't actually leave until we have all the inbound processing pods secured to the drum and our antimatter reserves recharged. From my discussions with Cecilia and David, that will be within a week. We're looking forward to participating in colonial activities until then." His eyes sparkled with extreme excitement. "Actually, Nikko's arranged for me to take a turn piloting a watercraft for a team of biologists before we go!" He grasped his nautical mug in anticipation as smiles beamed around the table.

Elke smiled thoughtfully. "What a perfect ending to your time here, Captain. We'll be sad to see you leave. Alright, anything else? No? Very good. Cecilia and David, I need a moment of your time. Everyone else, let's get to work." George glanced over at his colleagues, but they refused to meet his eyes. He looked over at Elke, but her eyes had gone cold, betraying nothing.

As everyone left the room, Elke stood and walked over to David and Cecilia, who glanced at each other and also stood. As Elke approached, their height disparity became more apparent. She stopped half a meter from them, looked down, then sat on the edge of the table to look them in the eyes.

"This morning, I received an urgent message from my superiors. That message made it very clear to me how badly various systems need the metals you've found in this system, and instructed me...INSTRUCTED...*ME*..." A never-before-seen fire rose up in her eyes; her insult was apparent in her raised voice. She paused, and cleared her throat. "Not 'recommending', mind you, not 'suggesting', but 'instructing' that I do whatever I must to allow you two to establish mining operations on Eden."

Cecilia's face went white. Whether with fear or discomfort, Elke couldn't say. She stuttered, "C-Captain, may I say..."

Elke raised a hand and silenced her. "First, it's no longer 'Captain', it's 'Governor'. Second, let me make a few things extremely clear to you both.

I do *not* like receiving 'instructions', especially when I am being 'instructed' to do something I've already decided to do. Yes, that's right, I've already decided that these metals are essential and that we need to find a way to safely extract them. And, frankly, I'm also adequately motivated by my share of the profits from this venture."

She looked at them pointedly. "I want to be abundantly clear that this will not happen again. You will come to me and reason with me. You will *not* attempt to go over my head again." She stood, straightened her uniform, and looked down at them, like a disappointed mother staring at her scolded children.

"Do I make myself *clear*?"

David averted his eyes, "yes...Governor." Cecilia stared straight ahead, tight-lipped and still pale.

Elke nodded, "excellent. You will prepare your proposals for extracting these metals with minimal, and recoverable, environmental impact. You may want to explore, with George, whether there's an opportunity to take advantage of this 'succession' phenomenon to aid environmental recovery. You will present these proposals to me, my staff, the Xenobiology team, and the colonial leadership at next week's meeting. Your proposals will satisfy all of them, or they will *not* be approved. Dismissed." David and Cecilia quickly exited the conference room. As they did, Elke thought she saw George swiftly walk out of sight at the far end of the hall, but she couldn't be sure.

▪ DAY 9

George approached David and Cecilia, smiled, and raised a hand in greeting. "Do you two want to go first again?"

David shrugged, "sure. I'm guessing you have some amazing life forms to share, so we'll review the boring rocks quickly and yield the floor."

George laughed nervously. "Well, I certainly think they're interesting, but that's why I entered this field. But I find the rocks compelling, too!"

Elke approached them and asked, "ready? We have a good crowd; I think it's time to start."

They nodded, and Elke turned to the crowd. The room's seats were filled, and a few stragglers were finding places to stand in the back. She noticed that some of the colonists were adjusting their wardrobes; apparently, local retailers were making and stocking what had become fashionable while they were in flight. She made a mental note to investigate. As Governor, perhaps she should start dressing as a professional civilian rather than a deep space officer.

George, David, and Cecilia made their way to the raised platform. She could see her senior officers sitting together in the front row on the right. Jian sat with them. Her series on relativistic travel had been well received and, now, she needed new fodder.

Elke cleared her throat. As she began speaking, other conversations ceased.

"Welcome! I am certainly looking forward to this update from the plane-tology and xenobiology teams. However, I'm going to ask for your indulgence as we get through a few pieces of business."

She looked across the crowd and found Jenna. "Jenna, would you please join me?"

As Jenna approached, Elke continued, "it's been almost three weeks, now, since we reached the system and began our approach to the colonial platform orbiting Eden, and a bit more than a week since we moved from the *Lucky*

Strike to the colonial frame. My staff has been impressed with your resilience and stoicism. We want to thank you for your support and cooperation."

Jenna said, "thank you, Governor. On behalf of the colony, we want to express our appreciation for the patience and professionalism of your team." She held out her hand, palm up, and swept her arm across the senior staff sitting before her. "Please join me in acknowledging Moira, Nikko, John, and Dani." The crowd applauded, and the four officers turned to smile and wave in acknowledgement. Elke paused the celebration with a raise of her hand, "Sri, we also thank you for the crossing, and we'll be sorry to see you leave. It's a bit more than a week now, right?"

Sri stood, arms outstretched. "It's been an absolute privilege and delight, but yes, duty calls and we must take our leave in just about five days."

"I heard you get to fulfill one of your dreams before you go?" Elke asked rhetorically.

Sri laughed. "Yes. Nikko's arranged an opportunity to pilot a watercraft on an Eden ocean tomorrow. As you know, my love for the ocean has been unrequited; thanks to Nikko, I shall finally have my chance to fulfill a lifelong dream." He bowed towards Nikko with an exaggerated flourish, generating light laughter. "And then, we'll be finalizing our departure preparations." He smiled and sat down.

"You will most definitely be missed," Elke commented quietly. "Well," she raised her voice again, "I'd offer the rest of my staff an opportunity to say a few words, but I really want to hear from the scientists. David, did I understand you'll be talking first?"

"Yes," David answered as he stood, "but I'll be brief. I know everyone really wants to hear from George, not me." David smiled and winked as he turned to Cecilia and said, "let's start with the asteroid mines."

He flicked his wrist computer and a projection of the asteroid belt appeared. He and Cecilia walked through the asteroids, and as David spoke, he slowly spread his hands apart, expanding the asteroids until one large body appeared directly before him.

"This body was identified by the deep space drones as our first extraction opportunity. It's some 370 kilometers in diameter, and we believe it, like many of the asteroids, was part of a planet that came apart, stressed by the tidal forces of the larger gas giant. Those tidal stresses led to concentrations of the minerals that interest us, so the deep space probes..." He raised his hands and pulled them apart again, "built their first mine here." As the asteroid grew, the straight and curved lines of an artificial structure became clearer, until the mining apparatus was as large as David. There were five mushroom shaped structures interconnected via pipes of varying diameter, surrounded by processing pods. Some of the pods were tethered to the asteroid, and one was slowly towed out of the projection view by a drone.

"You can see," Cecilia commented, "that we've already collected and pulverized quite a bit of high yield ore. That ore is now being processed in these pods, which are on their way here to be attached to the *Lucky Strike*. We now have seven mining operations like this one operating among the asteroids."

David flicked his wrist computer again, and the asteroids were replaced by a rocky hillside with a rivulet of water running down the hill and into a stream. "We've found a number of promising sites thus far. Here we have a placer deposit, near the tsunami site that Esteban was exploring a few days ago, searching for the source rock. If it's really promising, we take drill cores to build a model of the ore body. This one looks rich, but we're waiting for sample processing."

"And how do you know where to look?"

"An excellent question. We flew drones over areas that appeared to be geo-logically stressed, then sent landing parties to take samples." He smiled and added, "you can apply all the technology you want, but in the end, there's no substitute for a planetologist tripping over a pretty rock." Light laughter from the crowd encouraged him.

"So, to summarize: our deep space probes have been busy mining the materials used to build this colonial framework. We've also found asteroids with the kinds of stresses that isolate desirable materials and now have seven mining operations underway. And we have found some promising deposits on Eden, but it's too soon to know how they compare with what we've found among the asteroids. If you have any more questions, I'd be happy to answer, but I'm also anxious to hear what George and his team have found." After a few minutes answering questions, David tapped his wrist computer and the projection disappeared.

George stood, paused, and began:

"Here's the first bit of news: everything we've found to date affirms com-plete biochemical incompatibility between Eden and Earth. Now, we're not taking our environmental suits off yet, but everything we've found to date continues to build the case that humans will be able to safely walk the face of Eden. Local pathogens should not be able to infect us."

There were whoops and cheers all around. Even the other scientists joined in the celebration. George turned to see Elke clapping lightly with a bright smile on her face. They both knew what this could really mean for the future colony. Elated, but on task, George raised a hand and calmed the crowd.

"Of course, that also means that nothing that grows on Eden will nourish us. If we establish a colony on the surface, we'll need to grow our crops in soil that has been sterilized and cultured with Earthly microorganisms. So, while we would be able to walk the surface of Eden, we'd need to farm under domes." He smiled, "but, I must say, it's a beautiful world and well worth the effort.

Eden is proving to be about as different from Earth as it could be. You may recall, in our last briefing, discussing Earth's four nucleic acid base pairs, and how every world we've found that bears life uses different nucleic acids from Earth and each other.

Eden doesn't just have different base pairs than everywhere else; Eden has 5 distinct sets of nucleic acids!" He paused, then joked, "ok, I had a bet with Marita that you wouldn't find that as amazing as we do without an explanation." He smiled and pointed at Marita Park, "first round is on you!" Laughter trickled through the crowd; after it died down, he continued. "So, all life on Earth uses the same four base pairs, so we believe all life on Earth evolved from a common origin, called LUCA, or the Last Universal Common Ancestor. Here, there are five distinct sets of nucleic acids, so we think life arose on Eden five separate times. Three of those nucleic acid sets are relatively rare, and two of them seem to be the most successful. By the way, it's not just nucleic acids: the complex carbohydrates, the proteins, all of this organic chemistry in 5 distinct sets.

Now, you might ask whether this is unusual. I'd respond that we have explored two complex ecosystems that are completely different from each other, and, until we find a third, we can't say which one is 'unusual'.

But it gets even more interesting! Here, for example," he tapped his wrist computer, "is a very large organism one of our teams encountered in a rainforest." A projection of the herbivore encountered a few days before appeared in the center of the room. Murmurs erupted from the astonished crowd. Many stood to improve their view. A few momentarily recoiled from the strangeness of the first truly alien life form they'd ever encountered, but their curiosity got the better of them.

George gave the crowd a few moments to collect themselves, then continued. "We've collected a number of organisms, and we have abstained from naming anything. Well, some things we've nicknamed to simplify our discussions, but nothing will be formally named until we work out what's related to what and build a taxonomy. We just aren't there yet. So far, we know that there are five distinct life lineages on the planet, and each has plants, animals, and microorganisms. On Earth, we had only one lineage that taxonomists organized earthly life into kingdoms: an animal kingdom, a plant kingdom, and so on. Here, we have five 'super-kingdoms', and we'll be breaking that down further as we collect samples, analyze genetic material, and so forth. Well, back to this fellow." He pointed to the body, running his finger along the line separating the darker upper surface from the lighter lower surface.

"See this line? This line separates the upper third of the body from the lower two thirds. Now, every animal you've seen from Earth, including us, has 'bilateral symmetry'. A right and a left side that roughly match." He raised his hand and rotated it a few times while grimacing. "Well, almost every animal you've seen. There are a few things, like sponges and jellyfish, that don't have any symmetry at all, but, for the most part, almost any animal the average person would think about has bilateral symmetry." He pointed at the projection again, "but not this fellow. He has trilateral symmetry."

He tapped his wrist computer, and the alien creature began walking, its eyes scanning its surroundings. He ran his finger along its side and said, "the body is structured in three sections: a top, or dorsal segment, with right and left lower segments. The right and left lower segments each have three limbs. The upper segment, which has a thicker hide, also has three limbs. This first one," he pointed at the upper limb, "has a sensory cluster that's highly maneuverable, allowing our friend to focus its attention in any direction. Makes them hard to catch. Each of the six digits ends in a light sensor analogous to our eyes, but with a completely different structure. The digits also have chemoreceptors along their length, like our taste or smell. These fan-like projections are sound receptors like our ears. Those are the sensory capabilities we share. Here, along the edges of the dorsal segment, is a line of pits containing pressure sensors roughly similar to the lateral line on Earthly fish; it helps our friend sense air movement in its vicinity."

He continued, "note that I continue to say 'it' not he or she; we haven't figured out how to tell the boys from the girls, or even whether there *are* males or females quite yet. We do find that the dorsal segment has two other limbs that are held folded within protective sheaths in the back of the organism. We think they are intromittent organs, analogous to a penis, but we're not completely sure yet nor have we figured out how they'd work. They are connected to organs that might be gonads: sex organs. We have noticed that every specimen we've collected has the same structures, so they're either all the same sex or they're hermaphroditic or we're completely wrong about what the structures are for. We have collected a number of smaller organisms and are keeping them under observation in enclosures in our lab; we're hoping they'll share their secrets with us sooner rather than later."

He flicked his wrist computer again, and the organism became translucent. "See here, all the major organs are found along the core, where the three segments join. Note that there is a tube from the mouth on one end that has a variety of structures in the main body cavity before ending in something like an anus. That's almost certainly a digestive system. We see one very large and flexible organ that's part of the tube with a number of other organs connecting to it, but no differentiation between a 'stomach' and 'intestines' like we're used to. Note that there is a circulatory system with a set of pumps rather than one heart like we have. We believe this is the nervous system; the cells look roughly similar to our nerve cells and they branch from these three interconnected structures that, we think, collectively comprise the brain. Gas exchange systems are here in these three sacks like our lungs. There is a supporting structure that is clearly a skeleton; but, of course, it's completely different from ours. We have one spinal column; our friend here has three 'spinal columns', one along each of the trilateral segments, with cross-members like our ribs between them to form a cage protecting the internal organs. Pretty strong, actually. Breathing depends on contractions of this diaphragm-like muscle at the base of the cage because the cage itself is more rigid than ours. Note the limbs and how they're connected to the body, completely different from the hips and shoulders of Earth animals. There are similarities. They're structures evolved to address the same challenges; supporting an active creature moving through an environment. And, cross-referencing with other specimens we've collected, you can see how these structures evolved from supports for fins, then legs that propelled an animal along on its belly, then letting this fellow lift himself above the substrate for greater speed and agility. You can see similarities with what we know from Earth, yet completely different."

"The mouth is interesting." He walked to the front end, tapping on his wrist computer, and the conical structure opened into three matching sections. As it did so, three, fleshy projections came out of the structure. "The mouth is a beak in three sections, with three tongues that have rasping surfaces and lots of chemoreceptors: taste buds. This fellow is an herbivore, using these tongues to rasp material from plants; the beak can be used to break plant structures to expose desirable food surfaces. In fact, one of our teams observed our friend here feeding in a fascinating way. It knocked down a small tree, then everted its stomach; uh, turned its stomach inside out, like a starfish would, to digest and absorb the tree's leaves. Fascinating!"

He stood and tapped his computer, and the organism disappeared. He said, "okay, a few things. I know I made a big deal about how whatever we find on Eden will not be 'animals' and 'plants' because they'll be entirely different from anything on Earth, but I'll also share with you that, amongst ourselves, we got tired of saying 'plant equivalents' and 'animal equivalents.'" Laughter rippled through the crowd once again. "So…we're calling the things that move through the environment 'animals' and the things that are green and growing 'plants.'"

As he spoke, a sequence of smaller projections began to cycle through: animals, plants, some aquatic, some terrestrial, some large with internal skeletons, others small and with hard exoskeletons. "The biodiversity here is simply stunning, and we've only begun to collect samples. We've only begun to discern which things are related to each other. We are continuously sending down landing parties to explore different ecosystems. We're collecting specimens and figuring out how they work mechanically and chemically. We've already identified an incredible number of novel organic compounds, though only time will tell which have commercial value. We're building collections of smaller living things in our labs to observe behavior as best we can. Absolutely fascinating!"

He tapped his wrist computer, and a set of icons appeared in front of him. "We've been building a library of what we've collected. You can use these icons to bring up projections like the one I've shared with you today. We're expanding that library every day with new species and additional options, like making them move, demonstrating behaviors we've observed, or revealing internal structure."

He tapped his computer, ending the show, and returned to his seat. The crowd clapped in thanks as he took his seat, and Cecilia clapped him on the back as he sat down. He breathed a sigh of relief. He loved this enthusiasm. They were falling in love with Eden, and God willing, their love for their new world would help him preserve it.

Elke stood and said, "that was absolutely *amazing*." She turned to the crowd, "thank you for coming. I'm gratified that so many of you find this as fascinating as I do. I'm very much looking forward to our next briefing. Any questions? No? Ok, have a good evening."

She turned towards George and shook her head slowly. "George, you know, I've been through most of known space. I've been to most of our colonies. I've seen more worlds and more kinds of worlds than, well, just about anybody. But this? It's one thing to talk about seeing aliens, but to actually *see* one? To see something so completely, utterly different?" She shook her head again, then said, "thank you."

He blushed and smiled. "We've *hardly* begun. We've collected from a few terrestrial equatorial ecosystems, but we haven't explored the temperate areas, arctic areas, chaparrals, or desert areas. We've collected some aquatic forms from the tsunami aftermath, but we haven't explored any aquatic environments directly."

Cecilia, David, and Elke's senior staff joined them as they spoke. Nikko asked, "I believe we have a boat deployment planned for tomorrow; we're letting Sri pilot one of our rigid inflatable boats exploring some of the tropical coastline waters. When do you think you'll be ready for land vehicles?"

George asked, "how would we do that? Would we use landing craft to deploy vehicles, then bring them back to the colonial hangar decks? Can we organize extended trips?"

David chimed in, "do any of the vehicles have controlled environments for getting out of environmental suits?"

Moira shook her head. "No, none of the current surface vehicles are designed for extended expeditions. Our landing craft will take them to the surface with you and stand ready to retrieve them as needed. Sri's boat is already in a landing craft hold. When the Governor's ready, we can fabricate shelters that you can set up and tear down as you progress, and you could pack replacement life support packs. That would still limit you; until and unless we build surface settlements, we'd need to recharge the packs here at the colony."

Elke pinched her chin, deep in thought. "Perhaps..." She sighed and put her hands on her hips. "Perhaps it's time to plan for extended stays, and perhaps it's time for you to build that observation blind. George, have you seen any signs of dangerous wildlife? I know the predators can't digest human tissues, but it's of small comfort to someone that's been killed by an Eden lion or bear to know they can't be digested."

George paused before saying, "we've seen signs of large animals but nothing yet that would be a concern. But I'm not ready to say they don't exist. We have to make sure the structures can protect the occupants and that landing craft remain ready to evacuate as needed."

"We already deploy drones," Moira interjected, "to monitor the area surrounding landing parties. We could easily build some sensors that would alert you to anything large in your vicinity and equip you with some weapons."

Elke nodded, "I want to see a plan. How quickly can you pull together a plan?"

"Give me a few days and I'll prepare something for you all to review."

"Very good. Okay, to the pub?"

▪DAY 10

The boat, a red, 10-meter rigid inflatable with 3.5-meter beam, was cruising a few kilometers from shore, slapping against small waves with a strong wake fanning out behind them. The pilot's station was partially enclosed with a structure that extended backward over the aft deck, providing shelter and storage for the scientists. One of them stood, braced against a console, and the other five sat along the inflated sides of the boat. They dug their feet into foot straps and rode the waves, warm within their environmental suits, nonetheless wincing when the ocean spray hit their faceplates. There were clouds in the distance, flat-bottomed towering masses of pale gray with golden highlights from the morning sun. The forecast was clear, but the landing craft was ready to fetch them should the weather change abruptly.

Sri was smiling broadly at the helm. Finally, after all these years, he was piloting a boat at sea. John, the official pilot from Nikko's team, was standing ready should Sri need help. But Sri had it well under control, and Sri's delight was so infectious John was happy to indulge his dream.

They'd been cruising over a rocky reef for some time, with depths averaging around one hundred meters. Sri consulted his instruments again and could see that they were approaching a pinnacle that would rise to within 10 meters of the surface. Beyond that, the bottom dropped back to one hundred meters for a short distance, and then true ocean depths loomed.

Sylvia Ogilvy, an oceanographer having her first experience with a real ocean, was standing in the middle of the aft deck watching a monitor linked with a submarine drone cruising below the boat. The drone's sensors were relaying images of the sea floor and various aquatic life forms passing by. She could see a change; the reef was gradually rising beneath them. The drone's sensors relayed extensive movement ahead in an area within fifty meters of the surface. She called out, "Sri, we're picking up something ahead. Can we slow?"

As they watched on their heads-up displays, they could see a large structure rising from a crest of the rocky reef. It was roughly conical, inverted, narrowing towards the bottom where it was fixed to the rock: a big, irregular cup about five meters wide at the top and some four meters tall. Myriad, colorful creatures of the alien ocean swam around the rocky structure. Judging by their colors, shapes, and varying sizes, there were a variety of species, likely a mixed representation of the super-kingdoms.

At the console, Sylvia asked, "would you like an array of the species?"

Ben Murali, an evolutionist, shouted, "please!"

Sylvia tapped at the screen, and each of them saw through their heads-up displays a collage of still images. The aquatic creatures' bodies were generally torpedo-shaped, designed for efficient movement through water, but some were designed for sustained cruising and others for brief bursts of speed. Some were thicker than others, some more laterally compressed. A few had odd, specialized shapes. At the front of each was either the conical beaked snout or the vertically oriented mouths they'd observed on the land animals.

The ones with conical snouts had trilateral symmetry and three sets of three paddles along the body. The first fins were a short, jointed limb ending in a sensory cluster on the top, and the remaining two in slots on the back. The ones with vertically oriented mouths were bilaterally symmetrical, with only two paddles on each side and eyes on either side of the head.

"Anybody going diving?" Sri asked. "It's about forty meters deep below the bow."

Ben, lost in the display, didn't respond right away. He caught onto the question and laughed, "absolutely!"

He reached under his seat for his fins, snapping them to the boots of his environmental suit. Sylvia laughed at him, "now, wait a minute! There's no way that anyone else but the oceanographer will be the first in Eden's ocean!"

Ben yielded. "We will gladly follow you down, Sylvia."

Four others began attaching their fins.

Sri worked at the pilot's console and the engine stopped. He pressed the anchor release, and they could hear the anchor drop from the bow. John walked to the stern and dropped the other anchor, then pulled on the line until it was taught and tied it to an aft cleat. The boat rocked gently as John dropped a small float on another line and watched it slowly play out behind the boat. Finally, he lowered a ladder over the stern and into the water.

John buzzed the helmet intercom to grab the diver's attention. They paused to hear his briefing.

"Okay, the structure that caught your interest is a few meters in front of us and at about forty meters depth. We're over a ridge. The depth is reasonably consistent fore and aft, but the bottom drops abruptly on the port and starboard sides to depths beyond anything you should attempt without helium in your packs. We're facing a gentle current. You'll be swimming into it and, when you've completed your dive, it will help carry you back to the boat. If you miss the boat, don't worry. Don't try to fight the current; just swim laterally to the current line playing out behind the boat and use it to pull yourselves in. I'll set the drone to circle the boat a half a klick from us. I'll let you know if it reports any pickup in current or anything big enough to be a concern. Our sonar hasn't picked up anything bigger than a meter anywhere in the vicinity. Still, please be extremely careful!"

The divers nodded. They sat on opposite sides of the boat, facing inward, fins fixed to their boots.

John added, "I know you've all practiced this in shipboard pools and with simulators, but this is the real thing. I'll count to three and you'll lean back to fall into the water. You have to go at the same time to avoid landing on each other. If, for any reason, you don't go with the group, wait, and we'll get you in the water after. Once you're in, swim to the bow line before descending. Sylvia, you have the lead."

He paused, then asked, "ready?" All responded with an "ok" sign.

"On three, lean back. One, two, three, GO!"

The group of scientists leaned back, letting the mass of their life support packs pull them off the boat. As they fell, fins flew up and flashed in the sunlight as they splashed into the water. They surfaced with broad grins, tapped their heads with a fist to indicate that they were ready to proceed, and swam along the hull.

Sylvia waited until they were all at the bow. "Okay, just like we've practiced in the pool. Ben, you'll dive with me. Cassie with Hakeem, Sam you'll partner with Hank. Everyone ready?"

They indicated they were ready, and each said "submerge." Their suits gradually reduced air volume until they were slightly negatively buoyant. They slowly descended, in pairs, along the bow line to about ten meters, their environmental suits regulating their air volume to maintain neutral buoyancy. As they swam down, their life support packs adjusted the oxygen content of their breathing mix to manage their nitrogen absorption rates and extend their no-decompression time. They kept their mouths open, stretching their lower jaws side to side, flexing to clear their eustachian tubes and relieve pressure-induced ear pain. They leveled off, suspended weightlessly, and felt themselves moving side-to-side by the watery currents. Smiling broadly, they began swimming towards the conical shape surrounded by brightly colored sea life. Each held their hands close to their chests, fingers intertwined, elbows out, letting their legs and fins do the work.

Visibility was excellent; they could easily see twenty to thirty meters. They leveled off at thirty-five meters depth, close enough to get a good look at the cluster of marine life. As they slowly circled the structure, the creatures darted away but gradually came back after realizing the explorers were not a threat. A cacophony of brilliant colors, their density steadily increased as more and more joined the swirl of movement.

Most seemed to be feeding on small objects suspended in the water, randomly interrupting their leisurely movements to dart forward and snap at whatever interested them. Some had vertically oriented mouths; others had trilateral beaks with sharp points. Those with beaks sometimes snapped at morsels or extended their tongues to grasp prey and draw it into their mouths, or to rasp away as they fed. Other large creatures cruised around, and the divers wondered whether they were predators waiting for an opportunity.

They could see a myriad of shapes and colors encrusting the substrate below them and struggled to make out what was rock and what was life. Occasionally, they could see something small darting in and out of crevices; other things made slow, steady progress over the relatively vertical surfaces. They saw small structures darting out of openings, suggesting something hidden was feeding by filtering food particles from the water. Tiny colorful limbs waved, some with the motion of the water, some in opposition to it. They slowly began to realize that very little actual rock was exposed through a richly colored, variably textured riot of living things. They could see one of the trilaterally symmetric creatures with its beak open, tongues holding fast to rock, the digestive organ everted and pressing against something on the substrate.

As they moved through the water, their little helpers began to differentiate between life forms and rock structures. Their suit visors increasingly added flashing outlines helping them pick out organisms beneath them. After about twenty minutes, they heard John say, "the drone is picking up something large heading in our direction, about four meters in length. We're unable to determine whether it's aggressive, so I think it's best you call the dive. You've been at depth for an extended period; you should make intermediate stops at twenty and ten meters, then a safety stop at five meters. It's moving slowly, so you don't need to rush your ascent. I'll let you know if the creature accelerates."

They looked at each other, hesitating, reluctant to interrupt their discovery. Was it worth the risk? Sylvia, ever cautious, ended the debate.

"Acknowledged. Coming up."

Sylvia led the team towards the boat. They swam to the anchor line, gradually ascending twenty-two meters by the time they reached it. Sylvia looked about, not wanting to ascend any further. At twenty meters, the team stopped and consulted their heads-up displays. They watched the surrounding water for five agonizing minutes, then ascended to 10 meters to repeat the process. After another five minutes, they ascended to five meters and a countdown appeared on their heads-up displays for a 3-minute safety stop. They looked past each other, into the gloom, wondering what was moving towards them. John's voice rang in their ears, startling them all.

"The creature is still moving towards you; it's now about 120 meters away. I don't know what bursts of speed it might be capable of; you should get out of the water *now*."

Their heads-up displays indicated less than a minute to go, but they swam upwards anyway. Sylvia reached the ladder and shouted, "Cassie, go!"

Cassie removed her fins and handed them up to Sri, then climbed the ladder. John helped her step into the boat, and she turned to help.

"Hakeem, you're next. Then Sam, Hank, and Ben. Go!"

Hakeem, having already removed his fins, handed them to John and began climbing. As John stowed his fins, Sri helped him into the boat. "We need to move faster!"

Sylvia nodded. The others had already removed their fins and were handing them to Cassie and Hakeem. Sam climbed the ladder quickly; Hank had a little trouble. Ben was climbing when Sri yelled, "it's speeding up, get out NOW!"

Sylvia threw her fins onto the boat as Sri and John grabbed her arms and pulled her into the boat.

They stood around the aft deck, scanning the surrounding calm water, hoping to see the creature as it passed. John retrieved the current line and hauled up the aft anchor as they all held their breath. Finally, Sri let out a sigh. "Well, that was exciting! We'll continue on our original heading. Hopefully, it passed us, and you can get back in the water when we reach the pinnacle."

Sylvia opened her mouth to speak, then paled as she abruptly pointed. From the east, a meter or so below the surface, a large torpedo shape emerged from the deep. Moving quickly, it shot just below the boat, rocking the craft with a small but surprisingly powerful wake. They turned to watch as it continued away. The drone relayed images, and they all stepped away from the boat's edge to observe. A long, thick shape with a conical snout, three pairs of large paddle-shaped fins and a lobed tail swishing from side to side swam past and disappeared into the gloom of the Eden-blue water.

Ben whispered as if the creature could hear him. "Impressive. Four meters, I'd guess a few hundred kilos, maybe more. I'm glad we got out of the water when we did!"

Sri lifted the ladder out of the water, then walked to the pilot's console and retracted the anchor. He hit another button and the drone returned to its nest in the hull of the boat, hatch closing behind it. Suddenly the boat lurched as if it had hit a rock; everyone recovered their balance. Cassie pointed to the north as the shape moved away. "I think it hit the boat!"

Sri and Sylvia exchanged a worried look; they could hear John call in from the otherside of the boat. "Landing craft, we may need extraction. A large marine creature just hit the boat. How far away are you?"

"Ten minutes. On my way."

They waited nervously. Sri said, "I hope that's the end of it and we can continue on our way. I just don't want to take any chances."

Sylvia responded, "no, you're doing the right thing. We're probably fine but we can't know for sure."

The boat lurched *hard* to port, knocking them off their feet. They heard a harsh squeak, a rasping sound, then a crunching sound. Suddenly a part of the inflatable hull collapsed and water began to seep near the stern. They could see the beak snapping at the inflatable hull, followed by rasping tongues drawing across it, ripping up the surface. Another part of the hull collapsed, and they could see the beak snap down on the inner hull.

John cried out, "we need you NOW. Whatever it is, it's poked a hole in the hull!"

The pilot responded calmly, "moments away."

The landing craft slowly descended towards them. As they watched, a second tongue entered the boat. As the tongues withdrew, they rasped at the hull and enlarged the hole. The water flowed through at an increasing rate.

Sri, backing away, shouted, "*let's get to the bow!*"

They scrambled forward, watching the hole in the hull enlarge. The landing craft continued to descend until it was hovering over the water in front of the boat. The forward hatch lowered and the pilot, standing at the hatch, threw a line to John, who clipped it to the bow of the boat. The pilot hit a switch and the line tightened, then began pulling the boat into the hold of the landing raft.

As the stern of the boat came clear of the water, they could see the beast whose beak was snapping at the edges of the hole in the hull, extending its rasping tongues through and drawing them back across surfaces. The top of the long, muscular body was gray fading to white below; the paddle fins waving in the air. Suddenly, the beak released, and the body fell to the water with a tremendous splash.

Sri turned to Ben and said, "I'm glad you got out of the water in time. I hate filing reports!"

Ben gave a nervous laugh. "Hell of a way to end your first command!"

▪ DAY 11

Ecologist Oyana Bao strolled along the beach. Nine days ago, the tsunami swept the coastline, exchanging stranded marine life for the ecosystem that once thrived here. The first scavengers on site were the ones that never left. Massive, lumbering creatures designed to endure the tsunami slowly feasted on the carrion abandoned by the waves.

They were improbable creatures and the subject of much debate. Their presence directly after the tsunami and their slow, shuffling gait precluded any possibility that they'd arrived from somewhere else. They seemed adapted to environments regularly inundated by tsunamis, with thick hides and short, stout legs with powerful grasping feet that they apparently used to hold fast to rock outcroppings as the tsunami washed over them. The scientists speculated that they fattened on the carrion left by the tsunami, then scavenged as best they could until the next event.

Their lumbering disrupted soil no longer protected by plant growth, enabling spores deposited by the pioneers of the last tsunami to sprout. Nurtured by droppings and without competition from established plants, they grew rapidly while they had the chance. Oyana was collecting samples and cataloging them as she discovered new varieties along her trek.

Small scavengers also emerged from the soil. It wasn't yet clear to her which were new progeny, and which had simply been waiting, dormant, since the last tsunami. But they were undeniably prevalent and did not appear to have the means to travel to the impacted area from anywhere else. Some were amphibious creatures, ranging up the beach to varying degrees depending on their ability to persist out of the water. They initially fed on the carrion, then on the first wave of pioneers. Still others had arrived on the wind.

Absolutely fascinating.

She could imagine the debates, over the next few decades, or perhaps even lifetimes, about how to organize the life forms they were cataloging into taxonomies. Did they all start at about the same time? Were the earliest the most prevalent? Were there, once, more than 5 super-kingdoms? But until then, their focus was on collecting samples, especially during the succession following the tsunami, because many of these pioneer species would only be found during the succession. And, just as they could already see how the lumbering scavengers were altering their environment, nurturing the next phase of the succession, she was anxious to see how the pioneers arriving now would alter the landscape in ways that would be exploited by new arrivals.

She was interrupted in her reverie when Gustav Bang suddenly radioed in. "Oyana, come over here, will you? We found something interesting."

Oyana looked right and left, frustrated that with radio headsets, you can hear each other but you can't tell direction. She responded, "where are you? I don't see you."

Then she saw Bharat Kwan a few meters up the beach waving at her. She joined him and Gustav. "What have you found?"

They were standing by one of the lumbering creatures she'd just been musing about. It was slowly feasting on a large carcass that looked like it had once been an aquatic creature; it was hard to tell from what little was left. Oyana presumed it smelled awful. The creature's beak was splayed, and the rasping tongues were slowly scraping flesh from the carcass and drawing them into the mouth. It was ignoring them completely.

Gustav pointed at the flank between the second and third legs on the animal's left side and said, "we've noticed these new structures on this individual and two of the others a few kilometers from here." Oyana leaned in and saw a field of small protrusions from the animal's flesh. They were about the size of a human finger, with a hard shell protruding from the flesh much like a fingernail from its cuticle. As she leaned further in, she could see movement inside a few of them. She deactivated her little helper. It wasn't helping to identify these novel creatures, and the flashing was distracting.

"Are you thinking, maybe a parasite?"

Bharat shrugged his shoulders. "At this point anything is possible. We're thinking about taking a sample, but I don't know how our friend will feel about it. If it causes any pain, we might get squashed."

Oyana chuckled, "well we don't want that. But look over here; what is this thing doing?" She was pointing at one of the protrusions higher on the creature's flank, and there was a smaller creature nibbling on one of the protrusions. It didn't seem to be getting any reaction. "I think these must be parasites, and I think this creature is feeding on them. Suggests to me that our massive friend won't mind if we take a few samples. Step back a few meters; I'll see whether I can take one."

They stepped back, and Oyana opened her kit to remove a pair of forceps and a specimen container. She gently grasped one of the shells and began pulling, gently then firmly. It resisted, then popped loose. A bit of fluid seeped from where it had been pulled, and then stopped. She repeated the process, placing five samples in the container before stepping away to join Bharat and Gustav.

"No reaction. Let's make a note of that. I'm looking forward to analyzing these samples when we get back."

▪ DAY 12

David, Estaban, and Terry followed a stream bed through a temperate forest in the northern hemisphere. The streambed rose on a gentle slope towards distant snow clad mountains. It was winter in the northern hemisphere. Their environmental suits maintained a comfortable temperature for them, yet somehow, they could sense the crispness of the air around them.

The terrain was uneven. The stream wandered through rocky outcroppings, running over stones smoothed by eons of its passage. Steep banks suggested intense springtime runoff when the snowpacks melted. Large logs wedged well above the stream's current level, and near the top of the banks, in the sturdier trees, clusters of dried litter from upstream plants lay stripped from their origins and abandoned here.

They'd found another promising placer deposit in a calmed area where the stream widened, allowing sediments rich with precious metals to settle. They were seeking the source rock.

They continued to climb, and the slope became steeper. As they slowly ascended the foothills of the mountains, snow patches were larger and more frequent. They could see more and more of the valley behind them. Tall trees formed a carpet of deep green stretching to the horizon, unbroken save the riparian zone along the streambed.

As they looked back, they could see small creatures venturing out of the forest to take water now that they had passed. It was interesting to see how they drank. The ones with beaks positioned their three tongues in a tube; the ones with the vertically oriented mouths put the lower edge of their mouths in the water. David made a mental note to ask one of the xenobiologists how that worked.

They could see smaller aquatic organisms swimming in the larger, deeper parts of the stream, aligned against the flow of the water. Esteban, who'd loved fly fishing in the colonial parks of his childhood, presumed they were hoping for morsels carried by the stream. He wondered what it would be like to stand in that stream and cast a line, how cold the water would feel, or what strange creatures might take his bait.

Terry paused periodically to examine the rocks, looking for insights into the movement of water through the layers of materials. There was much to study, so much to understand about how minerals were leached from rocks to feed placer deposits and how weathering patterns exposed ore bodies. She hoped her insights would help find promising deposits.

Walking became more challenging. They occasionally used their hands to help them climb when the flow of the stream cascaded over rocks to drop to a pool below. Esteban stumbled at one point, but quickly recovered. He seemed to favor his left foot, but only for a few steps.

From time to time, they scrambled across loose talus, the shards indicating breakdown of a rockface, likely from the seasonal expansions of freezing water. David picked at the shards, smiling at exposed minerals; his heads-up display picked out and highlighted flecks of precious metals in the rock, lighting up his visor with a multitude of colors.

The short day was ending as they came to a massive rock outcropping looming above them. They could see where trees had sprouted in cracks, their growth forcing the rock apart. They collected samples at the outcropping, and David smiled again. He could see telltale minerals and stressed rock formations suggesting that this was the place. He turned to look at the wilderness from their higher vantage point, struck by the beauty of the unspoiled wilderness.

They continued to work as the sun set and the moons rose. Tonight, one of the smaller moons was a thin crescent, the other small moon not visible. One of the larger moons waxed nearly full, and the other waned. The combined moonlight illuminated the surrounding forests and obscured all but the brightest stars. David had always harbored a secret passion for poetry and found himself musing about ways to adequately describe the beauty that surrounded them. He wondered whether the others truly appreciated this place, or laughed at his literary attempts.

He sighed.

He tossed a small piece of ore up and caught it, considering the challenges of drilling such a location. He'd have to talk with Cecilia about it. But he had a *very* good feeling that this was the place; this would be their first extraction project.

▪ DAY 13

Jian sat at her desk, finalizing her latest piece on the Eden expeditions.

Her audience, her now interstellar audience, was delighted with her series about the new colony and the lives of deep space officers and crew. Last she'd looked, she had over half a billion subscribers. She'd known this was the opportunity of a lifetime, but *half a billion* subscribers?

She knew her audience was anxiously awaiting her series about the amazing life forms the xenobiologists were collecting from the planet's surface. Still, she thought she should include some variety. She thought to include some human-interest stories, like the arrival of the first baby to the Eden colony (appropriately named 'Eve'), or the plans to establish a proper pub in the first ring, with spectacular views of the planet surface and decorated with Eden artifacts (properly quarantined, of course).

Earlier that day, she had interviewed Moira about the future of the colony. Moira explained that Eden would be growing at an accelerated rate, with an inbound ship carrying 1,000 colonists every year.

"How will you deal with that?"

They were sitting at a table in one of the command ring break rooms. Moira sighed and stirred her coffee. "It'll be challenging. The orbiting platform, as is, has substantial expansion capacity. And the expanded agricultural space built pre-arrival by the autonomous probes should support our needs for the first eight to ten years. The grasses we've planted are growing well, and we'll even have space for some cattle when they arrive in a couple of years."

Jian laughed, "I can't wait! I've never cared for goat's milk!"

Moira chuckled and countered, "well, it makes good cheese and yogurt. But, yes, I prefer cow's milk in my coffee. I don't care for this simulated stuff, but it beats goat's milk. Plus, the agricultural rings will allow us to raise crops that don't deal well with gravitational reorientation."

George entered the break room while they were talking. He selected a breakfast pastry and joined them as Moira leaned forward and tapped the table with her manicured forefinger. "The challenge? The challenge will be subsequent growth. We'll need to decide within two years whether we're building another orbiting platform or a surface settlement. Either way, we'll need to start construction of the space elevator as soon as possible. Not easy, you'll want to report on that process!" She went on to explain that the lead engineer for that project would probably arrive on the next colony ship. He or she would need to supervise building a base station and initial stalks, all with materials shuttled down on landing craft, until the space elevator itself could be used to send down raw materials harvested elsewhere in the system to feed their fabricators.

Jian turned to George. "What are the odds that we won't need domes?"

Moira leaned back to listen as George paused to collect his thoughts. Finally, he said, "honestly, I don't know. The xenobiology team has been discussing it, believe me! We see a breathable atmosphere and total organo-chemical incompatibility. We've put lab animals in enclosures exchanging air with Eden organisms, with no ill effects observed. At the moment, I can't articulate a good reason why you can't take a stroll on Eden right now without an environmental suit, except that we've only been here two weeks and I'm very conservative!"

Moira nodded, then added, "I'm conservative too. We all are. But, at some point, assuming we don't come up with a good reason why not, we're going to need to have some volunteers live on the surface…with constant monitoring and a quarantine process, of course."

"Agreed," George shrugged, "although it makes me uncomfortable. We're not asking for volunteers just yet, but I think that will happen soon. And there's something else we should be thinking about," he added. "Eden's not like the other planets we've colonized. Eden has a complex ecosystem. We should be thinking about the implications of that."

Jian looked at him, then said, "please, say more."

George leaned forward and meshed his fingers. "The other planets we've colonized have either not had indigenous life, or, if life was present, it was a quite simple ecosystem consisting of simple, one-celled organisms analogous to Earth bacteria. Domed human colonies, even when they reach optimal populations in the millions, won't negatively impact those ecosystems."

He paused, then continued. "Eden has a rich, complex ecosystem. As long as we're living in orbit, we're isolated from it, so our impact will be minimal. But once we're living on the surface, domed or not, our habitations will displace Eden biomass. We'll sterilize soil, enrich it with Earthly microorganisms, and grow our crops. Any ecosystem has a finite carrying capacity; a limited amount of living matter it can support. As Eden's Earthly biomass grows, its native biomass will decrease."

He straightened in his chair, his face becoming more serious. "The human history of Earth was a gradual exchange of more and more human biomass for non-human biomass. As the human population grew, as we added human mass, we did so at the expense of other living things. And it wasn't just human biomass. It was the complementary biomass of our agriculture. Crops. Livestock. Our pets and ornamental plants. And the vermin that exploit our communities. Remember that, in the early twenty-first century, not only was the human population exploding, we were also lifting people out of poverty. Wonderful for individual humans, but a human living an industrialized life has ten times the environmental impact of a human living in abject poverty."

He paused and shook his head. "As these trends progressed, we drove species into extinction. And driving species into extinction slowly but surely simplified ecosystems. The carrying capacity of an ecosystem increases with complexity and decreases with simplicity. We all know the results. By the end of the twenty-first century, with the human population approaching ten billion, we'd caused mass extinctions and reduced the carrying capacity of the planet, resulting in famine compounded by climate change until yet another pandemic culled our population by a billion souls. We're damned lucky we didn't go extinct."

He looked at his small audience with pleading eyes. "Now I know we're a long, long way from anything like that here. But we must bear this legacy in mind as we settle Eden. It would be immoral to repeat our earthly mistakes."

There was a long pause. The ecological disasters and associated geopolitical upheavals of the twenty-first century were taught every child in every school throughout settled space; they were, after all, what launched the human interstellar diaspora.

Moira spoke, "it was my understanding that all of this was taken into consideration in the early phases of colonial planning. That was the reasoning behind setting a population maximum of fifty million."

George shrugged and said, "yes, and some of us urged caution and conservatism. We need to monitor our ecological impact. Maybe Eden can support more. Maybe less. Orbiting capacity won't be a factor. We need to keep our plans fluid as we grow, adjusting as we increasingly understand the implications of our footprint."

Jian smiled and scribbled furiously on her tablet. She had her next piece for her series.

■ DAY 14

The landing craft was descending through the atmosphere of the northern hemisphere. Elijah, a young pilot with a quick laugh and an easy smile, took great pleasure in his work, chatting incessantly with the groups of scientists he brought to and from the planet surface.

"So…what ecosystem are we heading to now?"

Oyana smiled and explained, "the coordinates we gave you will take us to the northern hemisphere to a zone between taiga and tundra."

Elijah glanced at her, laughed, and asked, "can you repeat that in terms I would understand?"

Oyana smiled again. "My apologies. We're approaching the higher latitudes of the northern hemisphere, where, this time of year, it's winter. We plan to go to the same latitudes of the southern hemisphere in a few days.

On Earth, 'taiga' refers to the massive coniferous forests of the northern hemisphere and 'tundra' to the ecosystem between the taiga and the poles. Tundra has permafrost, soil layers that never thaw, limits plant growth.

On Earth, in temperate climates, deciduous trees produce broad leaves during the summer and drop them in the fall. Coniferous trees produce needles that are less productive but are more water efficient, so they keep them year round. So, as you move towards the poles, you see deciduous trees yield to evergreens. But even evergreens can't deal with permafrost.

On Earth, taiga was only found in the northern hemisphere because those latitudes, in the southern hemisphere, were mostly ocean. Eden's continents have ample landmass in both the northern and southern hemisphere to support taiga, so it will be interesting to compare them."

Oyana shrugged, "here? Everything's different. We're presuming temperate zone trees have a strategy to deal with winter, but we haven't been here long enough to figure that out. And we're curious about the animals.

111

We haven't found anything analogous to hair or feathers yet, so, if there are animals, how do they deal with extreme temperatures?"

Elijah nodded, "thanks for the quick lesson! It looks like we're almost there."

The landing craft descended to a broad cloud layer, blindingly white as it reflected the oblique rays of the rising sun. They cruised over the clouds, relying on the landing craft's navigation systems to guide them to the desired coordinates.

Oyana, looking past the pilot, mused, "I hope the scout drone accurately mapped the terrain!"

Elijah laughed, turning slightly to say, "drones haven't made a mistake yet!" He sighed, then said, "it sure is beautiful, isn't it? Are there clouds like this on Earth?"

"I asked Yvette the same thing," Oyana chuckled, "she's our meteorologist. She said 'absolutely!' Remember what I said about the climate being like Earth? The physical environment is so similar to Earth; that's what makes comparing the life forms so fascinating, seeing where Eden's living systems' responses to the physical environment are similar and different."

"Is it strange that Eden is so much like Earth?"

"We've found two worlds with complex ecosystems. The two planets are very similar physically. We really can't say whether that's strange or not; maybe these conditions are a prerequisite to complex life…ask me after we find two or three more complex ecosystems!"

They were approaching the coordinates and descended slowly, cloud wisps passing with greater frequency until their view was uniformly white. Their view gradually darkened to a light gray as moisture began to accumulate on the view ports. They descended further until broad, dark green became increasingly visible through the thinning remaining cloud layer.

Finally, they were through the clouds. Dark green extended as far as they could see in all directions, save one. Before them, to the north, the green thinned, revealing growing white patches marred only by gray rocks as they progressed.

Elijah whistled. "Beautiful!"

Oyana laughed and asked, "enjoy your work?"

Elijah, with a broad smile, nodded vigorously. "I most certainly do!" He looked briefly towards Oyana, then returned his attention to his instruments. "Have you seen any of the oceans? Everything's beautiful, but, my god, the oceans!" He shook his head, "the most beautiful thing I've ever seen. Well, next to my babies." His eyes glazed over with joy. He was the proud father of a pair of twins born during the crossing, and was ready to share images whenever the opportunity presented itself.

Oyan patted his shoulder, "I think you'll be landing soon, yes?"

Elijah nodded.

The landing craft gradually slowed, finally reaching a stable hover a meter above the snow. Tall, dark green trees were visible through the viewports. Elijah unstrapped, then stood and raised his voice. "ALEX, WAKE UP! Time to launch the hovercraft!"

Alex, the tall and slender hovercraft pilot, was slumped in his seat, legs outstretched, and sound asleep. Hearing his name brought him around; he glanced about a bit, remembered where he was, and slowly stood. He stretched, yawned, nodded, and headed to the airlock without a word. He donned his environmental suit, went through the airlock, and descended to the hovercraft in the cargo bay.

The hovercraft was, essentially, a larger passenger edition of the suspensor pallets. It used the same antigravity suspension technology that the landing craft employed to hover without disturbing the surface beneath them. Elijah, having long wondered how this suspension technology worked, tried asking one of Moira's engineers to explain it. He didn't understand a word she'd said. Something to do with 'gravity shielding'. All he'd gleaned from the conversation was that it required a great deal of energy.

The hovercraft was a broad, gray platform, three meters wide and ten meters long, designed to deploy from a landing craft cargo hold. The pilot's position was in the center front with three rows of four four seats followed by a wide, flat cargo bed equipped with a gantry crane. A wide arc of transparent material afforded some protection from wind; but given that they were all clad in their environmental suits, further protection from the elements was forfeited to facilitate quick entry and exit.

As Elijah returned to his seat and prepared to open the cargo bay door, Oyana and her team donned their environmental suits. Then, in shifts, they passed through the airlock at the rear of the upper deck and descended the steps to the cargo hold. They boarded the hovercraft and strapped themselves in.

Alex satisfied himself that the team was ready and signaled Elijah. The cargo bay door opened. As Alex manipulated the controls, a drone and the hovercraft launched simultaneously out of the cargo bay to cross the lowered door to the snowpack. They headed east with the tree line to their right, their drone circling them to warn them of any large and potentially dangerous creatures. Behind them, after a moment, the landing craft ascended.

Alex asked, "Any idea how deep this snow is?"

Marsha, a planetologist, answered, "not yet, that's one of the things we'll be measuring as we go." She consulted her instrumentation and reported, "it's 1.2 meters deep right here. We'll need to fix snowshoes to our boots. The air temperature is -22° C, extremely cold."

Alex nodded, then asked, "where are we going?"

"Cruise along the boundary between the forest and the tundra," Oyana instructed, "we'll be stopping every one hundred meters for observations, measurements, and samples."

They continued eastward, stopping periodically for the team to deploy. At each stop, they walked northward across the snow, then circled the hovercraft, inspected the snowpack, rock outcroppings, and the forest.

After a few stops, Oyana spoke to her personal recorder. "Not seeing animal life at all. No movement on sensors. Strange. Diverse plant life though." She looked towards the trees, then observed, "the trees are in the silica structured group that we believe we'll find to be slow growing. We'll need to see whether that changes for less geologically stable taiga. We have encountered one surprise: nothing equivalent to conifer needles. The photosynthetic surfaces seem to be the tips of the branches. They're green and they end in widened blades with a waxy layer...I believe we'll find this conserves water. We'll need to examine them in the lab, but I don't see the easy break structures like the photosynthetic surfaces we found among temperate forest trees. That suggests they're not subject to browsing by any herbivores." She looked the other direction, out across the broad, bleak, snow-coated tundra. "The tundra rocks are intermittently coated with something crusty and organic; I suspect they are ecological equivalents of Earth lichens. We won't know which super-kingdom they belong to until we get samples to the lab." She sighed, then concluded, "we'll need to compare what we're finding to the southern hemisphere; I'm wondering whether the local animal life is dormant. Or maybe the lack of easy break structures means that there are no animals grazing these trees."

They continued, stopping periodically. Though they could not see the sun through the dense cloud layer, they could tell from the dimming light that the day (so short this far north) was drawing to a close.

Alex asked, "do you want to continue into the night?"

Marsha paused, "the temperature is dropping; it's -38° now."

Oyana read the room; no one was ready to leave. "Let's make a few more stops, then call Elijah to take us back to the bar."

The team cheered.

The expedition made three more stops before the drone alerted them to movement to the north. Alex turned to Oyana and said, "it's too dark for it to send us any visual, but infrared does show a heat signature. Want to go see what it is?"

Oyana didn't hesitate; this could be their first animal sighting. "Please!"

Everyone piled into the hovercraft. Alex worked the controls and the hovercraft turned north.

"We'll make this the last stop," Oyana said. "Let Elijah know we'll be ready for pickup soon."

Alex nodded.

They continued north, slowing as they approached where the drone had seen movement. The temperature continued to fall as the landscape darkened; it was now -45° C. It was difficult to see in the dim light, given that the overcast blocked the reflective light of the three moons that had now risen.

They slowed. Alex monitored data received from the drone as well as the hovercraft's scanners.

Movement.

He pointed to their left and turned the craft. In the dim light, they could only see that something fairly large was moving very slowly, too slowly to perceive without staring. Alex slowed to a stop when they were within twenty meters.

He asked, "do you want light?"

Oyana hesitated, then said, "sure. Let's turn on some light. But we can't know how it will react; let's stay aboard the hovercraft, ready to back away."

Alex worked the controls. The drone came to a halt five meters above the movement, then gradually illuminated the area. As the light grew, they began to make out what appeared to be one of the trilaterally symmetrical creatures. It was about three meters long, walking very, very slowly on six thick, short limbs with a thick, short tail. It ignored the light, and them, completely as it approached a large rocky outcropping.

After a few moments, the team left the hovercraft and encircled the creature. Its sensory cluster emerged briefly as they approached, held aloft on a very short arm emanating from the forward dorsal surface. The six eye stalks wavered as it scanned them, then the cluster returned to the fold in the back, presumably to conserve body heat.

As it neared the rocks, its short beak splayed. Three short tongues stretched out, exploring the outcropping. They stopped moving when they came to a larger encrustation of the material growing on the stones. The creature leaned a bit further in and began exerting its digestive organ, pressing it against the encrustation. The eversion was limited, presumably to conserve water and warmth. It stopped moving.

They watched for about fifteen minutes; the creature remained motionless.

Oyana consulted with her team, then said, "I want you all to walk back to the hovercraft and be ready to board. I'm going to see whether I can take some measurements from our friend here."

Marsha hesitated, "is that wise?"

"It seems very like the scavengers we found at the tsunami site," Oyana responded. "They're extremely tolerant. I'll be ready to run, though."

As the other backed away, Oyana opened her pack as she stepped forward. She removed a series of scanners and placed them along the body of the creature.

Alex asked, "what are you doing?"

"This is a portable version of the scanning technology we use in Sick Bay. It'll give us a reasonable scan of its internal structure. Won't hurt it a bit."

Oyana monitored her wrist computer, wary of the creature's movements, as she finished placing the sensors. It did not move at all. She then removed the sensors in the same order she'd deployed them, then walked back to the hovercraft.

"Alex, can you reduce the light to the minimum required, and have the drone record its movements, if any, while we wait for Elijah?"

Alex nodded and worked the controls.

Marsha asked, "what did the scanners show?"

Oyana turned to the group and said, "we'll need to review when we get back, but it seems to have the same internal structure and the tsunami scavengers. Trilateral skeleton. It's homeothermic, ah, warm blooded. Core body temperature is about 40° C, much like its cousins in other ecosystems. There's a very thick layer of fat under the skin; presumably that's what allows it to endure the temperatures. It doesn't seem to move much, so it doesn't need much energy. I'm wondering whether those encrustations are resistant to digestion; that may be why it's taking so long." She paused, "or, it's just very slow moving given the temperatures."

She looked back at the creature and mused, "I wonder what, if anything, preys on that?"

▪ DAY 15

Governor Elke Lubandi strode purposefully through the familiar, broad, gray, sparsely populated corridor on her way to what had once been her conference room.

The *Lucky Strike* reconfiguration was complete. The two command rings, providing work and living space for Captain Sri and his officers and crew, were now rotating alone. Adjacent to them were forty rings of processing and storage pods. They contained the results of years of mining and refining by the autonomous probes that initially explored the Eden solar system and built the colonial frame now supporting her bud of a colony.

As she ducked through the portal to enter, she felt an odd sense of nostalgia. Sure, this had been *her* conference room for nearly a decade ship's time But, as she glanced at a familiar stain on the wall from an unfortunate coffee spill one morning, the room and the ring were identical to the command ring constructed by the probes as part of Eden's colonial frame.

Truly, she mused, her attachment was to the memories and the people, not to nearly indistinguishable physical places.

Captain Sri and his senior staff stood as she entered. He smiled as he said, "Governor Lubandi, thank you for gracing our final staff meeting before departure."

She smiled and waved them to their seats, taking her place to the right of the Captain's chair. They'd delayed the *Lucky Strike's* departure as long as they could to give them as much opportunity as possible to share in the colony's discoveries, but it was time to go. She thought about teasing Sri about his boating misadventure, then thought better of it. Beginnings can be awkward times.

Sri turned to Amar and asked, "Amar, have you finished plotting our course?"

After a moment, Amar looked up. "Oh, yes. All plotted and entered. Quite similar, in fact, remarkably similar, to the outbound course. The relative movement of the two stars and the orbital positions of the two planets will result in a return total distance almost identical to the outbound journey. Assuming we leave on schedule, we'll arrive in fourteen years and, uh, sixty-eight days, which will be 764 days ship's time." He paused, then smiled awkwardly.

Elke turned to Sri and grasped his hand firmly.

"Godspeed, Captain Rizzo."

He held her hand, then released it. "Thank you, Governor Lubandi." After an awkward moment, Sri smiled. "Seems anti-climactic, doesn't it?"

Elke laughed lightly, "trust me, I've done this a number of times. It always does."

She turned to the rest of the room and her voice filled the room. "It was an absolute pleasure flying with you. I'm sorry you can't stay as we continue to explore Eden, but we'll keep you posted as you begin your acceleration until we can't exchange messages, which I think will be in a week."

The *Lucky Strike* officers came to attention and saluted her; she returned in kind.

"Alright, we're adjourned," Sri resolved. "Governor, I'll escort you to the hangar deck."

As they walked through the hall, he said, "thank you for not mentioning my boating misadventure."

She laughed and said, "it didn't seem like the time. And I doubt anything will be taking a bite out of the *Lucky Strike*."

"One can never tell."

The ship's drum was only configured for the current complement, so it didn't take long to reach the hangar deck. As they entered the hangar deck and approached Elke's shuttle, he continued, "I expect our staff meetings will stay short; there's only twenty-five of us."

"Yes, that's what I'd expect," Elke nodded in reply. "And they're good people. See if you can work on Amar's discipline, will you?"

He laughed, "no promises. You knew, when he signed on, what he was like, didn't you?"

She shrugged her shoulders. "He had an excellent reputation. And I thought I could make some progress with him." She paused, then smirked. "He's oblivious. Probably part of what makes him so good at his job."

Sri laughed quietly in agreement. He smiled awkwardly, looking back and forth between the ship and Elke.

"Permission to hug the governor?"

Elke's eyes went wide with surprise. She laughed and they briefly, awkwardly embraced.

Sri said, "thank you for this opportunity."

Elke shook her head and said, "you've earned it. You've a bright future ahead of you, Captain. I happen to know you've had some glowing performance reviews."

"Well, that's good to hear. Well…" He paused, "I suppose it's time. I'll operate the hangar bay."

Elke nodded and boarded her shuttle tug. As the door closed, Sri went to the operator's room to oversee her launch and departure.

▪DAY 16

Jenna was in her conference room, meeting with colonial business leaders. This wasn't an official meeting; Nikko was invited to the official meetings. She participated in small talk, smiling at stories of petty disagreements between tradespeople and their customers or competitors. She was relatively young, and this was her first colony start; but she was a quick study and had thoroughly researched how other colonies bootstrapped their economies. She understood that most, if not all, of these tensions with colonial standards and business operations conflicts would work themselves out without intervention.

She called the meeting to order, then went around the room asking for updates. Everything was on track. Rules were being observed where possible and circumvented where necessary; those that needed assistance made their requests and their needs were met.

Their economy was still a closed system; the *Lucky Strike* had just departed, loaded with their first exports. Those exports, and the ones to follow, would ship out on a roughly annual cadence. As additional colony ships arrived and then returned, the system would establish a very rich source of income for everyone at the table, but only after their first exports reached their markets in fifteen years.

She knew privatization of mining operations was important to everyone in attendance, and she let that discussion run its course. It was important, but not urgent; there was very little she could do to accelerate that process. Governor Lubandi's many years of service at relativistic speeds had given her investments many years to grow; she was immune to supplementary financial motivations. Privatization would progress at its own pace.

But she fancied herself a visionary leader and wanted to initiate discussions of other opportunities unique to Eden. She took advantage of a pause in the conversation.

"Okay, people, I want to talk tourism."

After a pause, Chang Volkova asked, "do you mean establishing tours of the system for our colonists?"

Jenna smiled but shook her head. "No, I mean tourism of Eden, for customers at other colonies."

"Generally speaking," Chang raised a skeptical eyebrow, "we haven't seen a successful interstellar tourism industry given the distances. What did you have in mind?"

"You've seen Eden's ecosystems during the last public briefing. Wait until you see this evening's briefing. I've heard a bit in the Governor's staff meetings. The planet's surface is incredible! So much more compelling than any colony's zoological or botanical gardens!

So, I'm envisioning avatars, I'm envisioning drones and/or people capturing the experience of Eden ecosystems, and making those experiences possible through simulations. I think it'll be a very rich business and available to us long before the *Lucky Strike* reaches her destination."

The room was silent for a calculated moment as attendees exchanged glances. "Agreed, it's an amazing opportunity. We can imagine people paying a premium for quality experiences. When do you think we'd be able to start?"

"Soon. We're starting to talk about settling the surface. I'm a little concerned that the xenobiologists are going to try to restrict access." Jenna glanced out at the hallway, her lips pursed with sudden displeasure. "They're a little too focused on conservation, if you ask me. But we may be able to get some recordings from some of the landing party members. I'm working on it."

"Can you think of a way to get any specimens?"

"What did you have in mind?"

"The next colonial expansion ship will be here in a year. I'm thinking there are private collectors that would pay a fortune for living organisms from Eden. Do you have any idea what they're willing to pay for rare animals and plants from Earth? Imagine what they'll pay to be the first to have something from Eden!"

He leaned in, "I'm thinking we should start planning for ways to collect and transport a few iconic Eden animals and plants. It won't be easy, but the payoff? Worth the effort…"

. . .

Esteban was one of the first to arrive. He walked to the raised platform in the front and stepped up to one of the chairs on the right. Cecilia, who was talking with one of her mining engineers, noticed and addressed him. "Good to see you. How was the landing party?"

Esteban smiled. She was a very attractive woman and he wished she was single. *Or at least more adventurous.* "It was very good," he said, "we were in the eastern mountain range of the main continent in the northern hemisphere. We found some promising placer deposits. We hiked up the streambed and found, or at least we believe we found, source rock. We took some surface samples. If they're as good as we think, we'll be doing some drill cores. How about you?"

Cecilia, either oblivious to Esteban's interest or ignoring it, simply nodded. "Good, very good. Natalia was just telling me about some of the assay results from the lab." She turned to Natalia and asked, "you weren't processing any of Esteban's samples already, were you?"

Natalia looked intently at Esteban. He knew that look. He never quite understood why some women were so drawn to him and others not at all, but he had definitely learned to recognize that look. He gazed into her green eyes and smiled. She held his gaze a moment, then finally responded to Cecilia:

"No, I haven't seen Esteban's samples yet. We're still processing some older samples. But I promise you'll get my full attention as soon as possible." She flashed a knowing smile at Esteban.

Cecilia laughed gently, raised her hands, and turned to look at the stream of people coming through the door. Seeing David approach, she interjected, "I, uh, need to talk with David. If you'll excuse me?"

She walked towards David, then glanced back. Natalia and Esteban were sitting together and talking. She was smiling, running her fingers through her hair with her head tilted slightly to the side. Her right leg crossed over her left and her right foot began to bounce slightly. He leaned forward and emphasized whatever points he was making with lingering touches on her kneecap. She also noticed Esteban occasionally, apparently absentmindedly, scratching his heel with his left hand. David smiled and nodded towards them. "Another conquest?"

Cecilia rolled her eyes and shook her head, laughing. "He is *notorious*. But then again Natalia's always been a little predatory. I would have expected them to have crossed paths before this."

David nodded absentmindedly as he noticed George and Elke walking towards them. After they exchanged greetings, Elke asked, "shall we maintain the tradition of starting with planetology?"

"Sure!" David smiled broadly. "We've found some extremely promising source rock above a particularly rich placer deposit of precious metals. Gold, platinum, palladium, rhodium, ruthenium, just amazing! If the surface samples are as good as I think they'll be, we should take some drill cores as soon as possible. We might have our first extraction project!"

George grimaced, "and where would this be?"

As David described the location, George's discomfort grew. One of the xenobiology teams completed an initial survey there just a few days before; the site was as rich and compelling biologically as it apparently was mineralogically. A more thorough expedition was in the queue a few weeks out. Now it sounded like he'd need to speed up that schedule.

As he contemplated his next words, Elke asked Cecilia what the drilling process would entail.

"Well, this will be our first Eden drilling operation! As David said, we'll take some surface samples as part of a site survey. Then we'll do some initial cores, probably a dozen or so, some 500-1,000 meters deep. That will give us a great deal of information about the rock formation that will help us fine tune the rest of our drilling."

"What will the environmental impact be?" George asked sourly, perhaps more so than he intended.

Cecilia paused. "Minimal. I'm working with Moira on a modified version of the rigs we used on the other bodies in the system. We can fly them in, then we'll need to clear and level some twenty to forty square meters at each drill site over the initial few hectares. The leveling will be the hard part; the site's on a hillside. In any case, that'll give us enough information to build a model and decide where else to drill."

George stared incredulously. "Hectares?! We're talking about a pristine riparian environment! If you start clearing and leveling up to forty square meters at a time, you'll destroy it all!"

126

David leaned back and raised his hands. "George, you know as well as I do that riparian environments are our most promising opportunities! What did you expect?"

Elke, noticing heads turning towards raised voices, decided to defuse the situation. She looked at George, "riparian? Is that significant?"

He looked up at Elke, thankful for the interruption and the opportunity to collect his thoughts. He took a deep breath, "yes, riparian. The movement of water creates a riparian ecosystem distinct from whatever surrounds it. The water itself creates a medium for aquatic creatures, plus there's extra water for the plants and animals. And the physical movement of the water changes the environment, creating a streambed, depositing material, flood zones, and so forth. Like on Earth, we expect riparian environments on Eden to have different animals and plants from the surrounding forest."

"Exactly!" David interrupted vigorously. "And that movement of water wears down exposed rock faces. When the water slows, sediment is deposited, especially sediment heavy with precious metals. Over time, as we survey the planet, we'll find other deposits. But right now? Our initial best bets are riparian environments with placer deposits that lead us to source rock!" He turned to George, his expression revealing some disdain. "George, you know this! For worlds with liquid water, such erosion is our best way to find promising ore bodies. On Eden, that means riparian ecosystems. I'm not sure what you want me to do here, George. We need these metals!"

George's ire was building. "No place else has riparian environments. No. Place. Else. And you want to, to, to *what*, exactly? Clear twenty, maybe forty square meters at a time? Destroy the rock face, level sections of it to support drill rigs? Permanently alter the stream bed?"

"And *how* exactly do you expect to fund your expeditions, George?" The group suddenly became aware of Jenna's presence when she spoke. "David is right; we need these metals. The whole planet's a treasure house. These metals, and the exotic organics you're finding, are what will feed us and fund your research!"

George extended his arms, palms up, and leaned forward exasperated. "So, what? We should just destroy the first complex ecology we've found? What gives us the right? We're going too fast! We don't know enough about the impact, or the planet's ability to recover! Why can't you extract your metals elsewhere in the system?"

"George, we've been round and round on this." David sighed, clearly tired of this conversation. "The metals are here, throughout the system, but only Eden has the tidal forces to concentrate the metals. Sure, we're finding stuff among the asteroids, and I'm sure we'll find more on the other worlds. But it won't be as concentrated, so it won't be as economical to extract. And it'll be harder to find without your precious riparian environments wearing down source rock to produce telltale placer deposits. Jenna's right, George. I'm sorry, but she's right. We need these metals to fund everything else. We're here for the metals, and for the organics. But the science…it's just a bonus."

"This is neither the time nor the place for this debate." Elke raised her voice slightly to quell the growing fire. Her tone was commanding and firm, but still kind. She appreciated their passion and dedication, even as it caused them to butt heads philosophically. But fighting amongst her senior staff was inappropriate and unproductive, especially in public. She wouldn't stand for it.

She turned to George and said, "we will not drill anything anywhere without reviewing the environmental impact. You'll get your say in that review." David opened his mouth to comment, but she silenced him with a look. "We will *also* extract the minerals. Responsibly. We will preserve the planet's ecosystems. If that requires some creativity, some invention, well then we'll figure that out. Moira will work with you to design the gentlest extraction methods possible."

A long pause hung in the air before David finally held out his hand. "George, we're in this together."

George looked at his hand, sighed, and grasped it. "You're right. I understand. It's just…mistakes are rarely forgiven a second time, you know?"

"And that is a completely understandable fear, George," Elke stepped in. "Thank you for your hard work and concerns. David, thank you." She straightened up and adjusted her uniform, "now, shall we continue the tradition of starting with planetology?"

George and David looked at each other. "Sure," George replied. David nodded.

"Well then, let's take our places."

As they walked, Cecilia leaned towards David and commented, "Esteban and Natalia have disappeared."

David sighed. "They're adults."

PART 3:
INFESTATION

•DAY 17, 04:32 PLATFORM TIME

Esteban woke up with a full bladder, idly scratching his ankle. Sighing, he stood, stretched, and stumbled to the lavatory. He began to realize something was amiss. He tilted his head to one side, then the other, rubbing the back of his neck as he walked. Slowly he realized he shouldn't be alone. It was coming back to him now.

It had been a very pleasant evening. Natalia was, to say the least, enthusiastic and creative. He smiled. *And nimble.* As he relieved himself, he remembered some of the high points of the evening and smirked. She had relatively long feet, and there was this one thing she did involving the backs of his knees…

He glanced back towards his bunk; he remembered her clothing strewn across the floor. They are all gone now. She must have made her way out while he was sleeping. *Too bad.* He would have enjoyed a morning encore. He smiled again. *I definitely will be seeing her again…*

He sighed as he finished, washed his hands, then began flexing and posing in the mirror. He'd always been reasonably fit, but for some reason he was getting progressively leaner. His muscle definition was definitely improving. Perhaps from all of the physical activity on the planet's surface? He'd done some strenuous hiking, up and down the hillsides looking for source rock, hauling tools and samples while wearing his environmental suit. But that was days ago; would that have made such a difference? He looked at the mirror again, liking the leanness. Natalia seemed to like it, too, as did a couple of other ladies these last few nights. Maybe that was the exercise reducing his body fat?

He made his way back to his bunk, hoping to get a bit more sleep before his alarm. He had some lab work to do that morning. Maybe he should go fetch his samples and assay results personally? *Maybe I'll run into Natalie that way.*

He considered this, then decided on a more direct approach. "Send Natalia a dinner invitation. Wait until midmorning to send." *That should do it*. He'd check back later to suggest a place. Not that there were many options given the size of the colony. Still, there were a few nice places. Maybe he should suggest dining here? Might save some time…

He realized he was scratching his ankle again. That was weird. "Lights, half strength," and the bunkroom lighting came on. There was redness where he'd been scratching, and a small, raised spot with an ominous black dot in the middle. But it didn't seem swollen. *Maybe it's just chafing from the environmental suit boots?* The undergarment booties seemed adequately padded, but he'd done a lot of hiking. He couldn't recall feeling any irritation at the time, though.

He shrugged, thinking he might have to get it looked at. That wouldn't be so bad; he found one of the nurses "interesting."

"Lights out." He put his head to the pillow, remembering some of last night's better moments, imagining the nurse in Natalia's place, and gradually fell back asleep.

■ ■ ■

Esteban woke up with a start. He was lying on his back, head on his pillow. He glanced at his console. He'd overslept, significantly. His brow furrowed as he did the math. *Almost 4 hours. Not good.*

He started to get up, but his body wasn't responding as it should. He felt weak, groggy. He ached. His leg ached, his belly, a few other places. *Weird.* Why was he aching?

Crap, did Natalia share a bug with me? He frowned. They'd been on the same ship with the same people for more than two years. Sure, there were 1,100 people between all of the colonists and crew, but bugs were rare in the outer colonies. And anything anyone brought aboard with them should have run through the limited population by now.

His ankle was itching, driving him crazy. He absentmindedly started to draw his foot back and reach towards it but found he just couldn't muster the energy. *Why am I so tired?*

"Lights, fifty percent," he said weakly, concerningly, and the lights gradually increased. With what seemed like enormous effort, he used his elbows to push his body further up onto his pillow, crushing it against the headboard. This simple move exhausted him. *What's happening to me?*

His head was raised enough to look down at his body. Most of his body was exposed; he was wearing his briefs and must have pushed his sheet down to his lower legs as he slept. As he looked, he realized he was much leaner. No, not just leaner, he'd lost muscle mass, enough that his skin was a little baggy. There was something more. He stared at his left thigh, where it ached. He could have sworn…

There it was again! Something was moving under the skin of his left thigh.

Something was *moving* under the skin of his left thigh!

He stared. It was a small movement. It was brief; it had stopped. He began to question whether he really saw it. He felt a tingly chill of fear as he gaped at his own leg. *Must have imagined it. Couldn't be.*

It moved again.

There was something *moving* under the skin of his damn leg!

"Lights, one hundred percent." Nothing happened. He said it again. *Why didn't something happen?*

He realized he hadn't said it loudly enough to be heard.

He tried again, as forcefully as he could muster. "Lights, full brightness." Nothing.

He yelled, as loudly as he could, "contact Sick Bay!" Nothing, no acknowledgment. He couldn't speak loudly enough. His words were barely a whisper.

He stared at his leg.

■ ■ ■

Estaban startled. He must have fallen asleep. He couldn't believe he'd fallen asleep. *How could I have fallen asleep?*

He looked towards his console to check the time, but found he couldn't turn his head. Only his eyes would move. He looked up at his bunk room ceiling, at the partially dimmed light panels. He looked to his left as far as possible and could see the wall. He looked to his right, but without being able to move his head, he couldn't quite see his water closet door.

At least he didn't feel the need to get to the lavatory. But wasn't that wrong? Shouldn't he need to relieve himself by now? How long had he been lying here? *This is really, really weird.*

He tried to call for help, but he couldn't muster the strength.

His ankle was itching *madly*. His leg was aching, badly. His belly, his arm, a lot of places, they all *hurt*.

He realized he was looking everywhere he could look by moving his eyes, but he wasn't looking down at his body. He knew he needed to look, but the prospect was terrifying.

He slowly turned his eyes downward.

His skin was even baggier. *Fuck!*

His eyes widened with terror as he saw more movement under the skin of his thigh. It was *bigger*. It was long and thin. It was at least ten centimeters long and it was moving under the skin of his damn leg.

A cold chill ran down his spine. Something else was crawling under the skin of his abdomen. He could see something long and thin moving under the skin of his belly, inching towards his crotch, under the elastic of his briefs.

He tried to move his hand to probe it. But his hand wouldn't move. *Fuck this…this can't be real.*

Something wriggled under the skin of his cheek, moving towards his ear.

Estaban began to cry, silently. He didn't have the strength to whimper.

■ ■ ■

He came to, slowly. He must have fallen asleep again. *How?*

His skin was so loose now. He could see the knobs of his joints. His knees were so bony! He could count his ribs. He couldn't move anything.

But he was breathing. He could see his chest rising and falling, gently, slowly. He was breathing.

He stared at his belly. It hurt, worse than before. There was a sharp, localized pain coming through the skin of his belly.

There was something *protruding through* the skin of his belly.

It was growing, slowly, so slowly, while he watched in horror; it stretched and tore the skin around it, and it *hurt*.

Fuck! He was in pain! Lots of strong, sharp, tearing pains…

The thing was a little lighter than his skin. He could see the boundary of his skin, and the outline of the thing just beneath: a smooth mass, almost glossy. It had a pointed end, sticking up out of his belly. Hot tears filled Estaban's eyes as it painfully stretched the flesh of his belly as it slowly, steadily emerged.

A second mass was tearing through the skin of his belly. It was steadily growing, too. And one on his leg. And there was something rising under his brief. If he wasn't in agony, he might have found it horribly funny to see something rising under his briefs that wasn't his own penis.

He was so exhausted. his eyes began to close but the escape of sleep was denied him by the pain. His existence was pure pain. Things were cutting and tearing him as they grew out of him. He started to cry again.

■ ■ ■

He woke up with a start. He felt faint, almost delirious. He couldn't recall where he was. He looked up at the ceiling. He tried to raise his head; his pillow was scrunched beneath his neck and very uncomfortable. He couldn't move at all. His skin hurt. The skin of his belly was burning, and he felt a wet sensation that he gradually realized must be blood trickling down his side.

He remembered. *Oh fuck.* He remembered.

He looked down at his belly. They were still there. A lot of them. Each eruption was about the size of his finger. Some he could see better than others; some, depending on the angle and the light, he could see through. Inside, there was movement. He could see…something moving: little movements like a little limb moving, bigger movements like whatever was in there was doing somersaults.

He could feel the creatures vibrating with life; he could feel his skin tear little by little whenever the things moved.

He was exhausted, horrified, terrified. He stared, trying to understand what these things were, what was happening to him, what would come next.

All he could do was stare.

■ ■ ■

Estaban blinked slowly and heavily. He couldn't believe he'd fallen asleep again. *Didn't I just wake up? How long has it been?* There was no way to tell. He knew the time was displayed nearby, but he couldn't see it without turning his head. And he couldn't turn his head.

Nothing was hurting. The hurting had stopped! *Oh thank god. Oh finally.* But he was suddenly thirsty. So, so thirsty…

He looked down. He saw them. There were a lot of them. They were about the size of his little finger. There must be twenty of them that he could see.

The end of one of them was moving. He could see something small and thin, darting out of the end of one of them.

He stared. He couldn't take his eyes off it. He needed to see what was happening.

The end of the thing was changing. The tip was disappearing, eroding away from the inside. Slowly, as the tip eroded, a different shape emerged. He realized it was a casing, of some sort. He could see something pointy sticking out of the exposed shell. The pointed extension splayed, then closed. When it splayed, long, flexible things like tentacles extended out and quickly withdrew back into the casing. It almost looked like tongues tasting the air.

He could see other casings starting to move. He could see the tips starting to part.

The first one was almost a centimeter shorter than the others now.

It was crawling out.

A conical tip pushed through the cracked casing, followed by a pair of claws, folded up and tucked under its body. A pair of legs pushed away the egg casing, then another pair of legs followed suit. There was a tail, no, *two* tails. The tails were curled upwards. As he stared, he noticed that the body was changing color. Waves of color started by the conical tip, then washed backwards to the tails. It was mesmerizing.

It was a little smaller than the casing, about the size of his thumb.

As he stared, from the top of the body, near the conical tip, another limb unfolded that ended with a little bulge. Tiny extensions flared out from the bulge, moving randomly, like little fingers waving.

Estaban realized that it was like the thing George had shown in one of the briefings. Except for those little claws; that was different.

It crawled out of the casing, crawled down to his belly, and then, with claws still folded, placed its mid legs against the casing and opened its conical end. It used what he now realized was a beak to bite the casing, and he could hear the casing softly crunch. The flexible tongues came out of the conical beak, reached out across the casing, and drew back. The casing was being consumed. It broke and fell to his belly. The little claws shot out and grabbed the casing, and the beak bit at the casing, breaking it up, and the tongues shot out and drew back across the casing. As it fed, waves of bright red and yellow ran across its flesh.

There were more of them now all across his belly and on his leg. They all followed the same pattern: crawling out of their casing, then consuming the casings.

The first one had finished consuming its casing. The limb on top, the one with the bulge, was moving around. It began to walk across his belly, away from the others, towards his face.

It stopped by his left nipple. His chest was wrinkled and sunken in from the loss of pectoral muscle. The top limb aimed the bulge forward; the digits pointing at his nipple. The claws that had been folded under its body slowly unfolded, reached forward, and grasped the flesh of his nipple.

It tore off his nipple.

He could feel his skin stretching, then tearing, pinched by the claws. His vision became blurry; he thought he was going to faint from the pain. The beak opened and the tongues reached out to draw in his torn nipple. Then it opened again and bit down hard, tearing at the loose skin and widening the hole to expose the withered muscle and bone beneath it. The tongues rasped across and through the fresh wound, lapping up the blood. He could feel the little tongues tear away at his remaining muscle, like tiny meat hooks. It pushed its beak through the hole in his skin and began painfully biting and tearing at the muscle deep under the skin. He could see and feel those awful rasping tongues, now tearing away little pieces of his flesh into its beak. Its claws let go of his skin, and he felt them pinch and pull as it dug deeper into his chest; only the flashing colors of the little animal's hindquarters were visible outside the hole in his chest.

He could see another, having finished its casing, walk across his belly towards the fingers of his right hand. The claws shot out, grasping the tip of his forefinger, and the beak crunched through his fingernail, bringing tears to his eyes and a new agony. As he watched, he could see the tongues lash out over his bleeding finger; but instead of rasping, they held fast. He could feel them gripping the skin of his fingertip, he could feel the claws gripping the next part of his finger, as the little creature splayed its beak toward the ruined fingertip. The pulses of color accelerated, brightened. He could see something coming out through the mouth, something gray, glistening with moisture. The gray mass pressed against the fingertip, and began to *burn*!

The tip of his finger was on fire. He desperately wanted to fling it away, but his hand and his arm weren't moving. They would not move, and it burned and burned and burned.

Oh god, fuck! He felt something pinching his penis and he looked at his briefs and he could see something moving under his briefs. One of those monsters was tearing at the side of his penis!

Oh my god this hurts it BURNS, oh FUCK that hurts make it stop please SOMEONE make it STOP...

He felt something on his neck, then on his chin. It was coming up the right side of his face. He could see the conical snout and the sensory bulge all aim at his eye.

The claws shot out and grasped his eyelid. The conical snout opened, and the last thing he saw was the tongues reaching for his eyeball, the gray mass pressing slowly outward against his eye...

He screamed, so quietly, oh so gently, so weakly, as the burning and the tearing and the rasping went on and on and on.

▪ DAY 18, 08:47 PLATFORM TIME

Oyana stepped down from the landing craft to start her shift at one of the five observation blinds. The landing craft would wait for the team members that were returning to the colony.

The tsunami site was three hours behind the orbiting platform, so dawn was breaking and three of Eden's moons were still visible in the growing light. She paused to appreciate them. One was full, the other two scimitars of differing thickness. One of the planetologists had explained to her, while gazing into her eyes over a delightful beverage, that each of the moons orbited Eden on its own schedule. He was building a model that would predict the mix of moon phases one would see on a given night and help them understand the tectonic stresses and tidal schedules driven by the interplay of their gravitational fields. That had been a lovely evening.

She sighed and resumed walking towards the blind.

Nikko's team set up the blinds at strategic locations along and above the coastline. They were sealed with an airlock and maintained life support for the xenobiologists so they could take a break from their environmental suits. They were stocked with spare life support packs and a supply of food and fresh water.

In their planning, they'd debated ways to camouflage the blinds; ultimately, they decided this was of limited value. None of Eden's life forms had any reason to fear humans, nor did any of Eden's life forms seem to notice them at all, at least so far. They settled on prefabricated boxy structures, four-meter square and 2.5-meter tall, with a flat roof, and broad, tall observation windows on each side. The windows allowed the observers to see out, but were opaque from the outside. The rest of the structure was camouflaged to match their surroundings. Next to each blind was a smaller, similarly camou-flaged utility shed for specimen containers.

They'd also established perimeters of sensors, one and two kilometers from the blinds, to detect anything large enough to be a threat. So far, nothing large came by except the lumbering scavengers that endured the tsunami.

Xenobiologists were putting in staggering twelve hour shifts. Grids had been mapped about each blind with a randomized schedule of grid sampling and observation surveys to quantify the changing populations of pioneers as succession progressed. She was amazed at the biodiversity of the site. Already, some of the early pioneering plants and animals were gone, having sprouted from whatever reproductive artifacts were left in the ground. They grew rap-idly, then reproduced and deposited their next generation in the soil before being consumed by the next wave of pioneers.

That next wave included plants that took a bit longer to sprout and grow, but were now well established. They studied these plants to see how they supplanted their predecessors. Were they outcompeting them for resources? Were they altering the chemistry of the soil? Were any of them the early arrivals of the next wave of pioneers?

That wave also included animals that arrived by land, air, and sea. They mercilessly consumed the first wave of pioneers struggling to reach maturity and reproduce, while avoiding being consumed by each other. Soon, pre-sumably, they would also face the depredations of the next waves of pioneers. Already they were observing a variety of animals about the size of cats and dogs, up to about twenty kilograms. There were bilaterally and trilaterally symmetrical species. The largest were voracious herbivores, aggressively stalked by predators as they grazed.

She'd been pondering how they would know when the succession was complete. They'd surveyed comparable, undisturbed coast lines, so they thought they knew what the site was like before the tsunami, but it was impossible to be certain. Time would tell.

She'd stashed some of her favorite snacks in the blind's food storage, then prepared to conduct her first sampling run of the day. She was about to enter the airlock when an alarm rang. She tapped her wrist computer and saw in her helmet display that one of the perimeter sensors detected a large animal approaching the blinds. She walked back to the blind's console for more information. A loud notification informed her of a conference call from the other blinds.

Several heads and shoulders appeared above the console.

Bharat asked, "do we have any information on the creature that's approaching?"

Hasani replied, "not yet. All we have at two klicks are motion sensors. We know it's big, over the 50kg threshold, but we don't know how big or anything else."

"Here's hoping it comes closer," Oyana smiled confidently. "The little beasties here are interesting, but I'm ready to see something new."

Geoff Singh, one of the junior ecologists on his first surface assignment, was not as confident or as comfortable as the others. "How long would it take to reach the next perimeter?"

"Well, much depends on what it is, of course," Hasani replied with his eyes glued to the incoming readings. "And whether it's heading directly here or on a tangent through the sensors, but probably we have ten minutes or so."

Oyana interjected, "I was about to start a sampling run but that's not enough time."

Bharat responded, "no, you should definitely wait. I know we're supposed to run them on schedule, but safety first. Well, what shall we talk about while we're waiting?"

Martin laughed and said, "There are advantages to being a botanist; I don't need to think about my study subjects moving about. In any case, we could play a few more hands of poker!"

Oyana tapped a control on the console and a virtual deck appeared in front of her. "Sure, I was going to start one after my sampling run. My deal, as I recall. Jacks or better to open." She tapped the control again and five virtual cards were dealt to each player.

They got through two hands before they were interrupted by an inner perimeter alarm.

The cards disappeared, replaced by an image of the approaching creature in the early morning light. It was from the trilaterally symmetrical super-kingdom, just over a meter in length and about fifty kilograms, a bit bigger than a large dog. The first pair of legs ended in curved talons, curled under as it walked on its knuckles. The remaining four legs were thick, grasping the ground as it moved steadily, stealthily forward, sensory cluster deployed. Most of its body was a dull tan, but its beak was nearly black, as was its sensory cluster.

It was heading directly for Oyana's blind.

As they watched, it lowered its body and slowed its pace. Oyana tapped a control panel and their perspective shifted to a wide, overhead view. They could see that a set of the large herbivores were grazing about 200 meters from Oyana's blind, and that the intruder was heading directly towards them.

Occasionally, one of the herbivores raised its sensory cluster as high as it could, lifting the forward part of its body to extend its reach. The six eyes splayed to observe the perimeter. Then the creature would lower itself to resume grazing, and one of its peers would interrupt its own feeding to assume sentry duty.

The intruder slowed, pulling its sensory cluster closer to its body as it approached. Holding its body as close to the ground as it could, the creature's colors slowly changed to match the background.

Hasani gasped. "Did you see that? It changed color!" Someone shushed him over the intercom; Hasani didn't catch who.

It was now within two body lengths of the herbivores.

When it was one body length away, it suddenly reached out with its forward limbs and launched itself with its four rear limbs. The herbivores nimbly turned and ran, but not quickly enough. The intruder grabbed one with its forelimbs, talons piercing the prey's flanks. As it landed with its remaining limbs, it bit into the poor grazer with its beak. It broke camouflage, returning to its resting, tan color.

The herbivore struggled and tried to free itself. As they watched, the intruder deployed its sensory cluster to examine its prize, eventually staring into the herbivore's own eyes. The predator opened its beak and extended its tongues. It ran its tongues over the body of the herbivore, rasping tissue from its body as its victim struggled and squealed.

This went on for more than a few minutes. The predator rasped its tongues over the struggling body of its prey, stripping off its skin. Then, still staring into its prey's eyes, it turned over the herbivore, extended its tongues over its body, grasping it firmly, and began pulling the animal into its beak.

The observation team watched as the prey was slowly pulled into the predator's mouth. The base of the beak expanded to accommodate the girth of the prey. Eventually, its struggles ceased, and the predator finished ingesting it. It raised its sensory cluster and scanned the area, then began moving slowly in the direction of the herbivores' flight, eventually crossing the one kilometer and two kilometer sensory perimeters.

Oyana whistled slowly. "Amazing!"

Hasani whooped, then added, "that's the first large predator we've seen! I mean, I assumed, uh I mean we presumed they must exist, but, damn, holy shit I can't believe we observed that!"

"Yep…well, that's all recorded," Oyana tried to keep her cool, but inside her chest her heart was beating very rapidly. "But I'm overdue for my sampling run. I'll be back, and then we can play another hand."

▪ 17:28

The little creature, uncomfortably hungry, walked along a rib. Waves of somber colors washed slowly over its flanks as it searched for more sustenance. It felt, through the lateral sensory pits along its sides, the vibrations of little movements in the distance. It could feel the distinctive electromagnetic pulses of its own kind, as its siblings dispersed looking for food. But it could no longer detect anything coming from the host. The host was dead, the meat almost gone.

It was time to go.

It had been one of the last to emerge; it needed to get away and find some food before it fell prey to a larger sibling.

The creature deployed its sensory cluster and waved its eyes around. Nothing looked right; nothing resonated with its instinctual programming. Its electromagnetic senses were very confused. There wasn't any clarity about direction, no clear north or south.

But it felt air movement through its lateral sensory pits and decided to follow the air. The creature walked along the rib until it curved downward; it withdrew its sensory cluster, leapt forward, and landed on a yielding surface. It redeployed its sensory cluster and walked along that surface until it reached an edge with a significant drop. It scanned left and right, then turned to the left and walked along the edge. Its sensory cluster continued to scan, following the air movement.

But the creature didn't see its larger sibling crouched motionlessly, carefully and effectively camouflaged, down to the texture of background. It never saw it at all, just a flash of motion as two claws snapped out and grasped it, one claw snapping shut over its beak and the other grasping the fleshy body between its second and third pairs of legs. Waves of colors flashed in fear and panic and pain, from reds to purples and yellows to blacks, faster and faster as the tiny creature struggled. But, without the ability to open its beak, it had no recourse. It could not fight back as its larger sibling pulled it closer and bit into its flank. Tongues snapped out, shredded its flesh, drawing tissue back into the beak. The beak snapped again, severing the tiny creature's sensory cluster. The world went dark; the world went silent and lost smells and tastes. Only the pressure of the claws were left, along with the electromagnetic patterns from its sibling happily feasting. That, and the awful pains as the tongues ripped, and ripped, and ripped.

▪ 22:42

As the landing craft rose from the beach, the pilot radioed, "I'll be on station when you're ready."

Sylvia, without thinking, waved, and then realized the pilot couldn't see her. "Uh, acknowledged, and thank you."

Ben crouched in the surf next to a drone. He stood and tapped at his wrist computer, looking at the drone's boot sequence as it scrolled up his visor. After a moment, the boot sequence was replaced by a live feed of the breaking surf. He flicked his wrist computer again, and turned to Sylvia, "you should have the drone's feed now."

"Yep, I see it," Sylvia responded. "Good to go?"

Ben flicked his wrist computer again and the drone moved forward into the surf. Soon the feed was relaying images from calm water beyond the break. He instructed it to patrol the water a kilometer away from their section of shoreline.

There were six of them on the beach, including two of Nikko's operations people setting up an equipment station. Cassie also crouched in the surf, placing a specimen in a sample container with a pair of forceps. Hakeem stood near her, looking out over the water. Sam and Hank stood together further down the beach and seemed to be making casual conversation as they admired the horizon.

Sylvia turned to look out, and she could see why they were staring. Eight hours ahead of the orbiting platform, the next morning's bright orange sun was crowned with crepuscular shafts of light. Partially visible, as it rose it extended towards them a long, bright, glowing finger across the smooth surface of the water. High cumulus clouds tinged with brilliant color gradually spread across the skies from the east to above their heads and beyond. A few darker, denser clouds towards the south released rain that didn't quite make it to the ocean surface before evaporating.

Here, at about the thirty-fifth southern parallel, it was summer, and the air was quite warm. But they were on an eastern shore, and Coriolis driven currents were drawing water up from polar regions. Off the coast, prevailing winds pulled up cold, deep water presumably rich with nutrients from a submarine canyon. On Earth, such a combination would result in a very rich ecosystem; the landing party was here to see what such a combination might produce on Eden.

As the sun rose and withdrew its colors from the horizon, Sylvia called to the team and asked, "are we ready to explore?"

Cassie yelled, "yes!" The others raised their hands, then trudged back from the surf. They stood in a circle and Sylvia continued, "great, me too. Let's buddy up the same way we've been diving. I presume everyone's life support is green?"

Everyone nodded.

"Excellent. You've all seen the maps; the bottom drops steadily from here to about twenty meters depth, then angles down more aggressively to about forty-five meters. It then drops precipitously to over 300 meters depth. We will make our way out to the precipitous drop and descend to *no more* than fifty meters, because this will be our first dive beyond no-decompression limits. So, I don't care what you see at fifty-one meters; you will *not* descend below fifty, clear? See something amazing? We'll come back tomorrow." Nods all around.

"Your packs have all been equipped with a helium cartridge. Your environmental suit will monitor your depth and adjust your gas mix as we descend, mixing in helium as needed. That will let us minimize our decompression schedules on our ascent; we'll do stops at forty, thirty, twenty, and ten meters before ascending. Your heads-up displays will guide you as to how long to stay at each depth. And remember, there's no harm staying at a stop longer than you need to, so always wait until your dive partner is ready to continue ascending. Okay, is anyone not ready?"

Each of them tapped their heads to indicate they were ready, so she picked up her fins and began walking towards the surf. They followed suit and joined her down at the water.

As they walked, they could feel the waves breaking gently around their shins, sometimes challenging their balance. They continued until the water was up to their abdomens. Sylvia stopped, crouched, and attached her fins to her boots. She stretched her arms forward, tucked her head, and leaned forward into the water, descending gently to swim half a meter above the sand.

As she did so, she could see hundreds of small, silvery blue creatures swirling about her that had not been visible from above. She turned her attention to the sand and saw random movements as tiny creatures buried themselves or darted across the sand.

The water seemed reasonably clear around her, but visibility was limited by the density of living material. Their preliminary samples indicated the water was teaming with microscopic organisms, and it looked to her like some of the silvery creatures swimming by were using their large, open mouths to feed on the plankton.

As they swam forward, they followed the bottom until dropped towards the canyon, stretching their jaws to relieve pressure differentials in their inner ears. They began seeing rocky formations in the sand. The formations became larger, more frequent, and increasingly served as anchors for underwater plants. Sylvia wondered if they were entering the beginning of a marine forest.

Some of this plant material had washed ashore, and preliminary analyses suggested it was quite different anatomically from the land-based plantlife. Sylvia found herself musing about how much longer it would take to get a comprehensive understanding of the diversity of Eden's marine life, of all of Eden's life for that matter. She sighed quickly at the thought of the workload ahead of her.

They continued swimming through the undersea vegetation, occasionally looking right and left to check on their dive partners. At some twenty meters depth, they passed through a thermocline to 12°C water. Their suit displays kept them apprised of their remaining breathing gas and current mix. At this depth, their life support packs were still enriching their oxygen level; but as they descended, they'd start seeing helium added to their mix to reduce nitrogen partial pressure.

The bottom dropped faster, sloping downwards. The underwater forest grew all around them, towering above them and teeming with life. A dizzying array of marine creatures swam or crawled through the foliage. Eyes and antennae were visible in crevices of color-encrusted rocks, while sand moved in the open spaces by hidden forces. Thankfully, neither they nor their patrolling drone had encountered anything large enough to be deemed a threat.

Visibility remained poor. Sylvia doubted she could see more than ten meters. She often saw clouds of tiny organisms surrounding the rock formations and wondered whether they were immature versions of the creatures swimming through the growth. *So much to learn!*

They continued descending, now passing thirty meters. The slope steepened as they approached the canyon. The bottom was now mostly rock. The light got dimmer as they descended, and their headlamps activated to illuminate their immediate surroundings.

As they passed forty meters, a short alert told them that they were approaching the limits of their no-decompression dive. Soon, it would no longer be possible to simply swim to the surface; they would be compelled to make decompression stops in order to safely ascend.

They emerged from the underwater jungle, swimming over broad swaths of ocean floor between rock outcroppings of varying dimensions. The bottom leveled briefly before continuing to drop. They observed that the marine life patrolling the edge of the forest was larger; a different community of creatures swam above the rocks, presumably ready to dart to safety.

Suddenly, they came upon a wall of swimming sea life. Thousands upon thousands of silvery blue torpedo-shaped creatures, five to ten centimeters in length, streamed slowly past them. The dive crew marveled at this river of living creatures, stretching from as far into the gloom as they could see. As they swam by, their bodies were briefly illuminated by the divers' headlamps, revealing dashes of shorter wavelength: reds, oranges, and yellows otherwise lost at this depth.

Sam observed, "there must be predators following them, but nothing big enough to alert our drone."

As they moved slowly through the school, the stream parted around them, then resumed behind them as if they'd never passed.

The bottom plunged to a rocky cliff, followed by a nearly vertical wall. They swam to the edge, signaled to each other that they were ready, and began to slowly descend. There was very little ambient light; even the longer blue and green wavelengths were reaching their limits. They hadn't seen any plant life of any kind for some time. Visibility was poor, only about five meters, presumably due to the rich nutrients they could feel rising from the depths beneath them. The swimming sea life was in the five to ten kilograms range, moving slowly but steadily past them. They reached their target depth and hovered, slowly turning around and staring into the gloom. It was disconcerting to consider just how much open water surrounded them: fifty meters to the surface, 250 meters beneath them, and all the way to the next continent in front of them.

Ben, his voice slightly distorted by the helium, said, "I'm glad the drone isn't picking up anything large."

Sylvia laughed nervously, "I know what you mean. Well, we've done it. Anyone feeling dopey from the pressure?"

Hakeem held out and rotated his hand, fingers splayed. "Surprisingly little. I supposed that's the helium in our breathing mix. Last time we passed thirty meters, I could read my heads-up display and I knew everything was okay, but I couldn't tell you what the data meant."

Cassie laughed, "ha! I know what you mean!" Her voice was naturally very high, so the helium-induced pitch caused several divers to chuckle.

"Normally, we'd expect pressure-related narcosis from the increased nitrogen partial pressure," Sylvia explained, "but our packs have been displacing nitrogen and, to some extent, oxygen, with helium. It's affecting our voices, but it will keep our heads clear, reducing the duration of our decompression stops. If we went deeper, it would reduce our oxygen toxicity risk. It's what we all expected based on our pressure chamber simulations, but it's quite another thing to actually experience it. For today, we'll just ascend per our plan. Tomorrow, we'll run some drills. After that, you guys can start taking samples as we go."

"Good!" Sam gave her two 'ok' signs. "We're collecting great video, but I've seen a number of things I'd like to examine in more detail back in the lab."

They began swimming slowly upwards along the rock face, staring at surfaces and into crevices. The rock surfaces were such a riot of colors and textures that they struggled to sort them into distinct life forms. As they swam over the top of the cliff, Sylvia stopped and turned her head sharply. Hank noticed and stopped as well.

"What is it?"

Sylvia hesitated, then said, "I thought I saw movement. I don't see anything now, must have been my imagination."

They hovered a few more moments, then Sylvia said, "we have to resume our ascent; staying at this depth will extend our decompression requirements."

They continued ascending along the slope, nearing the forest, approaching forty meters depth.

Suddenly, Sylvia heard a gasp and turned to see Sam moving swiftly backwards, arms outstretched, plummeting towards the depths.

"Sam! Stop! You mustn't descend!" She turned to follow him over the edge and into the canyon, but quickly realized he was descending faster than she could swim.

"Sam!"

"It's got me!"

"*What's* got you?"

"I don't know! Help, I can't break free!"

The light of his head lamp rapidly disappeared into the murky distance. Their drone began sending warnings that something large, something other than them, was moving nearby.

Sylvia's helmet sounded alarms about her rate of descent. She stopped, stretching her jaw, desperate to equalize and relieve her ear pain. Her heads-up display flashed two data points in bright red. Her depth was eighty-two meters, and she was nearing the point past which she would not have enough breathing gas to last through her decompression schedule.

"Sam, can you see what it is? Can you get free? Sam…SAM!" The absolute quiet alarmed her greatly. *How the fuck did it get past our drone sensor?? Where is he!*

As the others caught up with her, she turned to face them. "We have to turn back."

"What about Sam?!" Hank yelled.

"You were his dive partner. Are you still getting data from him? What's his depth?" Sylvia tried to suppress a rising sense of panic.

"No," Hank responded. "I'm not getting data at this point! Last I received, he'd passed 176 meters."

Sylvia and Hank floated in silence as Cassie, Ben, and Hakeem repeatedly screamed Sam's name. Sylvia looked into Hank's visor, but he didn't meet her gaze. He stared over her shoulder, into the void that swallowed their companion.

"There's nothing we can do." Sylvia bit her lip to keep her voice steady. She was the leader; this was not the time to break down. "Even if we could reach him, he doesn't have enough gas to reach the surface safely. He can't even safely ascend to a point where we can rescue him or replace his life support pack."

Hank, Cassie, Ben, and Hakeem didn't move. They were staring into the depths, unblinking. Ben made an abrupt, guttural noise and began swimming downward. Sylvia grabbed onto him with both hands, kicking furiously to pull him back.

"STOP!" she screamed. "We must resume our ascent NOW! He's gone. There's nothing we can do for him, and we must ascend NOW!"

She pushed Ben ahead of her and grabbed Hakeem's shoulder. Ben turned to look at her and paused, as if ready to challenge her. But he lowered his head, nodded, and began to swim up the canyon wall.

Sylvia could hear soft weeping, although she wasn't sure who it was. "I don't understand what happened…" Hank muttered in disbelief. "Anything that could have taken Sam should have been spotted by the drone!"

"You said you saw some motion in the rocks," Ben said, "perhaps there's something large that hides in the rocks."

Oh god, the rocks. "You're right." A fresh wave of fear rippled down Sylvia's spine. "We should move further away from the rocks!"

They moved away from the rock wall, away from the false comfort of swimming near a reference point. They swam further into the gloom where, intellectually, if not emotionally, it would be safer as long as the drone continued to monitor potential threats.

They paused, very uncomfortably, at seventy meters, then at sixty, then fifty, and on up. No one spoke as they followed the guidance of their heads-up displays. Their instruments warned them that they were depleting their oxygen as a consequence of reaching depths well beyond their plan. Sylvia trusted their instrumentation to provide a decompression schedule within the limits of their remaining oxygen supplies.

They reached the top of the canyon and continued to maintain their distance from the rocks. They ascended higher, now swimming about three meters above the substrate, doing their best to wind their way between rock formations rather than swimming over them. They continued to stop every ten meters.

Sylvia tapped at her wrist computer and consulted her gauges. "I, at least, went eighty-two meters. That's over nine atmospheres of pressure, and I was breathing hard, trying to reach Sam…we all were. So I went through a lot of gas. It's going to be close. We may need the operations people on the beach to bring us life support packs at some point as we ascend."

No one responded, but a slow and sad ripple of nods ran through the ring of swimmers.

"What do you think happened to Sam?" Cassie quietly asked the lingering question.

"Something grabbed him," Hank said, "something we haven't seen before, something the drone didn't pick up until it was too late. I was next to him, but

I…I was looking the other way…I swear I didn't see or hear *anything*. By the time I turned, Sam was…he was…already dropping fast."

"Whatever it is, it won't be able to pierce his suit!" Hakeem interjected with hope.

Ben paused. "I hate to say this," he responded grimly, "but that's probably unfortunate. We can't know how deep it took him, but he's already past the point of being able to survive an ascent to the surface."

They were all quiet.

Sylvia blinked furiously. "Dammit."

"I've known Sam a long time," Hank said. "He loved what he was doing. He knew the risks."

A long pause.

"Maybe this was foolish," Cassie's voice quivered. "Maybe we should be using the submarine."

Sylvia reassured her, "we will, when we go deeper. But the submarine is bulky; it's easier to collect samples in our environmental suits. We…weren't expecting anything big enough to be dangerous to be hiding in the rocks."

Cassie paused, then said, "I know, but…we lost Sam."

"There's nothing we can do about that except remember Sam and learn from this." Hank spoke through gritted teeth, but not in anger. "For now, let's focus on safely ascending; that's what Sam would have wanted."

They hovered in silence, as their tissues slowly released nitrogen. They anxiously scanned the gloom around them; they still had a long way to ascend.

▪ DAY 19, 01:37
PLATFORM TIME

It could no longer feel the electromagnetic pulses of the host; the meat source must be dead and nearly consumed. The others, its siblings, would be getting hungry. They would start hunting. It was time to get away.

The hungry creature held onto a small piece of the host that had come loose. It backed away, dragging the piece with its claws, using its tongues to keep taking meat from the remaining scrap. As it backed up, it came to an edge. Just as it realized this, it took just one step too many, and began to fall. The creature clutched its small piece of flesh. It was food, and food was not to be lost! But its feet could not stop from sliding, and as it fell, it protected its sensory cluster by retracting it. Red and yellow colors pulsed along its flank with fear; then it landed on its back, safe on a soft and yielding surface.

It twisted and used its tails to right itself. The creature felt, through its lateral sensory pits, a movement of air. It didn't trust its electromagnetic senses for direction, so it redeployed its sensory cluster and followed the air current backwards. Dragging the small piece of meat, the carnivore greedily consumed every last shred of sustenance as it backed away from the useless corpse and towards the movement of air. It could sense it: the chance to get away.

Finally, the piece of meat was gone, except for a hard, predominantly calcium piece with no more sustenance to offer. It drew the piece closer, biting it with its beak; it broke through the calcium exterior, then everted its digestive organ against the hole. It found nothing to absorb. It withdrew its digestive organ, released the small piece, folded its claws, and resumed following the movement of air.

At last, it came to a mesh covered opening, and crawled through to continue following the air.

▪ 07:05

Elke entered her conference room and found the attendees subdued. She realized she was right to reschedule their weekly staff meeting; they needed to deal with the loss of Sam right away.

She took her seat, looked at Nikko, and gently asked for a debrief.

Nikko cleared his throat and looked about the room, noting expressions. He began, "yesterday, Sylvia led our first technical dive." He looked about the room again, then continued: "The landing party had done all that could be done in pools and simulations. It was time to make our first dive beyond decompression limits. Sylvia planned a conservative dive."

Jian interrupted, "can you help me understand 'technical dive' and 'decompression limits'?"

John paused, then sighed. "Yeah, we're going to need to explain this to people. I'll send you some references on the physiology of diving. But here's…a summary. There are limits to how long a diver can stay at depth. Pressure increases with depth. When you breathe at higher pressures, your body absorbs nitrogen from the air you're breathing. The deeper you are, the higher the pressure, and the faster you absorb nitrogen. When we talk about 'decompression limits', we're talking about limiting our time at any depth so that we can return to the surface without absorbing enough nitrogen to cause decompression illness. Folks used to call that 'the bends.'"

John raised his eyebrows and looked over at Moira, who took over. "Uh yes, so, when we planned for Eden, we thought through how to adapt the environmental suits we use in the vacuum of space to maximize our working time underwater. Our life support packs recirculate the air we're breathing, replacing the oxygen we've consumed and removing the carbon dioxide from our exhalations. Under normal conditions, breathing in space or working on the surface, these packs last up to twelve hours, much longer than our usual working shifts. Nitrogen's our problem, so we modified our life support packs to adjust the gas we're breathing. At shallow depths, we increase the oxygen content to reduce the nitrogen content, which lets you stay underwater longer. As you descend, though, that same level of oxygen can be toxic. So for long, deep dives, we re-engineered the life support packs to start adding helium to the breathing mix to reduce the nitrogen level without subjecting the diver to toxic oxygen levels.

So, even with the sophistication of our environmental suit's life support, there are limits to how deep you can go and how long you can stay, without running out of no-decompression time. And once you do run out of no-decompression time, you have to plan, as you ascend, to stop and wait multiple times to let enough of the nitrogen you've absorbed to safely leave your tissues. We call this technical diving, because there's a lot of technical work calculating how much time is required at each depth. Our environmental suits do all the calculations, of course, including the added complexity of managing our breathing mix as we ascend to manage the rate at which we release nitrogen."

Jian sat back in her chair, "That's...a *lot* of complexity!"

"Actually," Nikko interjected, "we kept that pretty high level. But yes, you get the point. So, ultimately, our divers can explore the first thirty to forty meters of water without the complexity of technical diving and they can explore the next fifty or sixty meters with technical diving. After that, we have to use the submarine. It works, and we're thrilled to have it, but divers have so much more freedom of movement for their work."

"So...about Sam..."

"Yes, back to Sam. The team understood all this complexity intellectually, and they practiced everything they could in pools and in simulations. They've had a few no-decompression dives, as you know, and those went well.

Anyway, Sylvia knew this would be the first technical dive, so she kept it simple. There's an ecosystem of great interest in the southern hemisphere; a deep underwater canyon near shore. Seemed a perfect place to take the team to fifty meters depth and then ascend according to their decompression schedule." He paused, his eyes glazing over briefly with pain. "As...they began to ascend, something, something *large*, grabbed Sam and pulled him very deep. They did their best to follow him, aborting at eighty-two meters depth. The last data his dive partner received indicated he'd passed 176 meters and was descending rapidly at the time."

"But, why did they stop?" Jian leaned forward in her chair. "Based on what you said, their life support would have kept them going, they...could have gone after him, why didn't they go after him?"

"Because he was effectively dead already." John's voice cut across the table, catching Jian and Nikko off-guard. "First, we cannot know whether the creature that grabbed him had already killed him. Second, he was descending too rapidly for his life support system to effectively manage his gas mix. By that depth, if whatever grabbed him hadn't killed him, the oxygen level he was breathing probably did. Finally, he'd reached a depth at which his body was absorbing so much nitrogen that he wouldn't have enough breathing gas to survive his decompression schedule. His oxygen reserves would run out at a depth beyond any of Nikko's people's ability to save him."

A solemn pause hung in the air. Jian looked downward, then resumed eye contact. "I…I didn't understand that. I couldn't understand why they… abandoned him."

"It's not a simple thing to explain," Elke reassured her, "and I expect a number of our colonists are struggling with why they abandoned him. We'll need to provide an explanation." She looked to Nikko, "I would like you to craft a general statement citing references that people can consult until they understand as much as they need to."

Nikko nodded, "we're already working on it."

Elke smiled sadly, "I'm not surprised." She looked at David and asked, "how is Sylvia?"

He tapped the table uncomfortably, "she is…taking it very hard. She feels responsible, of course. But…" he looked at George briefly, "as it has been pointed out, they all understood the risks. I think she'll be ok. Getting back to work would probably be the best therapy."

"I agree," George said. "The team is already planning their next dive. I have a few volunteers to take Sam's place." He smiled, "and we're already picking a few organisms to formally name in his honor: a fitting tribute to the memory of a xenobiologist."

David smiled as well, "funny, we were thinking of naming that underwater canyon after him as well."

"I think those are fitting tributes."

Elke leaned forward, elbows on the table, with her chin resting on her meshed fingers. She closed her eyes for a moment, then opened them to say, "I think we all need to remember that these explorations are dangerous. Sam is the first loss of this expedition; let him be the *only loss*. I do not want another death on our watch." She looked at Moira and Nikko, "my understanding is that the drone didn't detect this, this *thing*, until it grabbed Sam and began descending. Do we have any additional information, and do we have any plans for corrective action?"

Nikko exhaled sharply. "We have very little information. Whatever it was, it apparently was motionless among the rocks until it attached Sam. It was large, approximately 400 kilos. We don't have anything about its shape, unfortunately."

"And there's not much we can do about the drone's ability to detect something that's not moving." Moira said this very matter-of-factly, but the lingering faces of displeasure caused her to change tone. "We will adjust the sensors for increased sensitivity, which, of course, increases the probabilities of false positives. We're also working on how to make it smarter about filtering false positives. But, generally, we believe the teams would rather have an occasional false alarm than miss a threat."

"Agreed," George said.

Elke leaned back in her chair. "We're running long here. Understandably, of course, but I have a topic that will require a bit of time. Before that, does anyone have anything pressing?"

No one responded; everything, apparently, was running smoothly.

"Very good. Well, we've been here almost three weeks. We've had numerous landing party expeditions with only one tragic outcome. It is my understanding that we have found ourselves to be completely incompatible biochemically with the indigenous life forms. In our last meeting, we tentatively agreed to establish an encampment without environmental suits at the one-month mark; that's in another week." She leaned forward and asked, "are we still comfortable with that schedule?"

George grimaced. "You know me, I remain very, very cautious about these things."

"I understand and respect that caution. Let me ask you this: what would it take for you to believe we are ready?"

He shrugged, "honestly? I don't know. Other than actually successfully establishing such an encampment, I don't know. And, even after a few weeks or months, I'll still worry. But, that's me. We all have different risk tolerances."

Elke looked at John and raised her eyebrows. "I agree," he said, "that we've established complete biochemical incompatibility. We *should* be able to live on the surface without succumbing to any indigenous pathogens, and without worrying as to whether we're contaminating Eden. This works both ways, right? If we're walking around without environmental suits, we're carrying Earthly microbes, but that biochemical incompatibility should prevent them from surviving without us." He paused, then continued. "But if you're asking me whether a month is long enough, well, I don't know. This is new to all of us. We don't have any prior experience to guide us."

Elke nodded, and rested her head back onto her hands again. "I think it's time to ask for volunteers. We need to know, and, as you've so aptly stated, we're at a loss to identify anything else we should research. Inform your teams about the level of risk, then ask for volunteers."

She leaned forward and continued, "pragmatically, we have to consider our protocol if, despite our efforts, something does infect them. We will do our best to cure them, of course, but then what? Do they spend their lives in quarantine? How do we determine whether it's safe for them to leave quarantine? Our volunteers have to understand that there is a possibility, a very remote possibility, that should something go wrong, they are risking their lives or could spend the rest of their lives in isolation."

She let that sit for a moment, then resumed. "The real question is what is the right number of initial volunteers. With all that in mind, I'm thinking two people to start, with a schedule to add additional people as time passes without incident. We will establish a camp for them: shelters, and perimeter security based on the predator observed from the tsunami site blinds. We can observe for a predetermined number of days and decide from there. Thoughts?"

John ran his hands through his hair anxiously. "Let's think of this along two lines. First, is this the right approach? Second, when should we start? As to the right approach, well, it's hard to argue with. No matter how much research we do, ultimately, it will come down to volunteers taking the risk. This seems to be the most reasonable way to approach the risk. As to the timing? Again, it's hard to argue; I can't come up with anything else to test in preparation for volunteers."

George drew in a long, deep breath; David and Elke eyed him curiously. Finally, he exhaled and dropped his shoulders. A little bit of anxiety seemed to fall off him. "I can't argue with John. Or you." He smiled, "and I freely admit I won't really believe it's safe until you have a thriving community on the surface. Even then, I'll have my fingers crossed."

Elke looked around the room, eyebrows raised. "Anyone else?"

After no one objected, she placed her hands on the table. "Great. This is a big decision. You've heard the proposal; ponder it over the upcoming week. Think about it, pick it apart in your minds, and, at our next staff meeting, if nobody has come up with an objection, we'll work through the process of soliciting volunteers. I will deliberate on minimum qualifications for the volunteers."

She stood up and surveyed her senior staff. Their faces were still somber, but she could see hope and a bit of excitement in their eyes. "Okay, we're done then. Let's get back to work, and let's be safe. Thank you all."

▪DAY 20, 03:47
PLATFORM TIME

The creature walked through the air ducts, following the air current. This environment was *so* strange. The terrain was flat and unyielding. There was no sun or moons, nor were there stars. It could not reconcile what its electromagnetic sensors and its gravimetric sensors were telling it; it was unconfident about any sense of direction. There was no green growth. All it could do was follow the movement of air. A neutral color slowly washed over its flanks, reflecting anxiety.

It held its sensory cluster high; the flexible eyes spread widely, scanning for any movement. It could feel vibrations through its lateral sensory pits, but nothing close. And it could feel the electromagnetic signatures of its siblings, far behind it. They were also following the air, some on the same path, while others decided to take different junctions and faded with distance.

The creature paused. It felt a different kind of movement, then a different kind of electromagnetic pulse. *Prey?* The colors along its flanks brightened slightly, moving a little more quickly. The eyed fingers of its sensory cluster aimed to the right, seeking more information.

There! It could see something moving in the distance, moving generally towards it. The flashes of color along its flanks brightened further and accelerated; then, as the prey came within optical range, the colored flashes stopped. It slowly changed color and texture to match its background and held still.

Very, very still.

The prey was a reddish-brown, with a generally flattened body and multiple limbs. As it watched, it observed that six of the limbs were used for walking. Two long, thin limbs at the front moved slowly, up and down, and side to side. *Sensory organs?* The prey was coming slowly, steadily closer. There were little, waving limbs at the front, and periodic bursts of movement.

One of the long, thin limbs touched something and the prey stopped. Then, as the two thin limbs touched the object, it walked towards it, lowering its little waving limbs to examine the item. As it watched, the little limbs touched, then tore at the obstacle. *Food!* The prey was *feeding*. The creature became increasingly excited; it wanted, so much, to flash bright colors. But it didn't. It held still, and slowly, ever so slowly, pulled in its sensory cluster close to its back, the eyes focused on the prey. Its ears waited to hear some sound of the creature feasting; the chemoreceptors waited to pick up some indications of what the prey would taste like.

The prey finished feeding and resumed moving, coming closer; the long, thin limbs waving side to side, and up and down. It watched the prey come closer, then stop as it felt the touch of one of the long thin limbs. The prey was still too far away, but almost, *almost*, close enough. The creature held very still and let the prey's long thin limbs continue touching it, not reacting. It was learning the vibration patterns from the prey's movement, the electromagnetic signature of its body.

The prey moved forward again, slowly, intermittently; apparently curious and unafraid, the prey started to move on by. Its left long, thin limb kept touch on the predator, as the right long, thin limb shifted to explore other surfaces.

The prey took another step, then another. *Almost. One more step…*

Pulling its sensory cluster close to its body, the hungry creature snapped out its claws and grasped the prey. Its left claw clamped around the front end, behind the mouth and in front of the forelegs; the right claw fastened onto the body behind the third pair of legs. As the claws closed onto the prey, its sensory receptors were flooded with information. The prey had a shell. The right claw was clamped on broad flat shapes on the prey's back that tried to move. *Wings!* The prey had folded wings. The lower part of the right claw was pressed against a series of plates, a body armor that slowly compressed under the strength of the right claw. There, *that's* where the prey's best parts would be: the densest meat, the fats, the organs.

It drew in its claws, pulling the prey towards it. Bright, bold colors quickly pulsed along its flanks; it no longer needed camouflage. The prey struggled, gripping the surface with its feet. But the prey could not resist the strength of its claws or the weight of its body as it relentlessly pulled.

As the prey came closer, it opened its beak, then bit into the flank of the prey, just past the third pair of legs. Body fluid seeped from the wound: a clear, viscous, sugary body fluid. The prey fought hard, its limbs pulling and pushing, trying to get away. The electromagnetic pulses strengthened; the prey was in pain, a lot of pain.

And that excited it very, very much.

It bit again, enlarging the hole. It extended its tongues. Their rough surfaces didn't get much tissue from the shell of the prey, so it used them to hold onto the shell around the hole and everted its digestive organ. It pressed its organ hard against the hole and into the prey. It tasted *wonderful*.

Slowly, it pulled the prey's body closer, pressing its digestive organ further and further into the shell of the roach. The hunter reveled in the flavors and the textures of the tissues it encountered; it thrived off the sensations of the prey's struggles as it tried to pull away, the electromagnetic pulses coming from it as it suffered.

It did its best to consume it slowly and carefully. Excited by its pain, enthralled by its agony, the predator did not want to kill its prey prematurely. It wanted to *relish* it.

Then, as the prey finally died, it settled into dismembering the carcass, patiently, methodically. It thoroughly extracted every bit of tissue that it could, and skillfully used its tongues to recover body fluids that had seeped from the body. It would not waste any of the sustenance.

Satisfying.

Using its lateral sensory pits to scan for more movement, the creature made its way down the air duct, in search of more prey.

▪ 09:08

The hovercraft moved swiftly over the dunes, heading towards a rocky outcropping. Like Earth, there were a variety of desert ecosystems. Water was always the limiting factor, but for some deserts, dryness was seasonal. Living things designed to endure protracted periods with little to no water flourished when the rains came. But this was a bleak desert; rainfall was exceedingly rare and very little life endured.

Still, it was beautiful, clean. The sky was untouched by clouds. The red rocks were particularly stunning in the early morning light; their color revealing intense mineral content; their shapes reflecting the work of sand-bearing winds. Islands amongst dunes that shifted like waves under the dry wind: a slow, reddish brown inland sea of sand and grit. Precious little green complemented the rocks; very little moved that wasn't pushed by the air.

The landing craft deployed the hovercraft within a few meters of a particularly large outcropping. The hovercraft approached the rock, and the planetological team stepped down onto the rock to start exploring as their pilot took the hovercraft to the far side of the outcropping to wait for them.

The rock was raw and challenging to traverse. They took surface samples as they progressed, grunting as they climbed up and down the surface. Very little was growing on the rock; very little organic matter accumulated in the cracks.

The rock outcroppings were volcanic pipes, just over a dozen of them, of various sizes, running roughly in a line east to west. Once raging volcanoes now long gone silent, they'd been weathered by rains that were now only a distant memory. Now winds and sandstorms continued the work, until only the hard, rugged, igneous pipes and dikes remained. Some were modest, and some were massive. One was large enough to create its own climate, capturing the rare clouds that did pass. It stole what little water the wispy clouds had to offer, and became a refuge for living things not fully adapted to the brutally arid climate around them. Terry was enthralled. Trying to understand how water was moving through the sedimentary rocks still clinging to those pipes was a scientific dream.

All of the pipes were mineralogically promising. Their samples would need to be tested, but Enrique and Sun were optimistic about them. They boarded their hovercraft, proceeded to the next rocks, and repeated the process.

As they approached the next outcropping, they circled and decided that their best access would be from the east. The hovercraft floated slowly over a relatively flat portion of the rock, but the adjacent formations were daunting.

Terry asked, "What's wrong?"

Enrique hesitated, then admitted, "oh, nothing, I'm just regretting not having climbing equipment. I would have liked to take some additional samples, but I didn't want to try those formations without proper equipment, especially in this environmental suit."

"Yeah, well, that was on Esteban's list. Not like him to miss an opportunity to visit the surface. He must have had another one of his...epic evenings!"

Enrique smirked and said, "I hope he has the decency to keep it to himself this time!"

Terry made a face, then sighed. "Well, there's certainly room on the hovercraft. I'll make a note to just keep some climbing equipment aboard so we don't have to think about it."

Enrique turned and said, "I noticed you staring at that scruffy little bush. Something interesting?"

Terry smiled sheepishly. "Not particularly. I don't know…everything is new, so I guess everything is interesting, right? Anyway, I took a few samples from the plant and a couple of the small, flying insect-like things for the xenobiology team. I don't know whether they've poked around any of the desert ecosystems yet. And they've been pretty good about bringing us rock samples, even if they don't know what to look for." She laughed, then said, "I had an odd thought as I looked at that shrub. Some of those little creatures will spend their entire lives on that one plant. It's their entire universe. And here I am, how many light-years from where I was born? How many light…I don't know, decades? Light-decades…is that a thing? That's a thing, right? From where I was born? Or from Earth?"

She shook her head, looked at him, and smiled. "It's just amazing, to think about how far we've come, how alone we are out here. How completely, utterly alone."

Enrique nodded slowly, "I hadn't thought about it quite that way, but, now that you mention it, yeah. Wow."

After a moment, he sat up and cleared his throat. "We uh, better get back to work."

The hovercraft lowered, maintaining a few centimeters clearance above the rock. They stepped out onto the rock and moved towards the face, ready to take more samples.

▪DAY 21, 13:55 PLATFORM TIME

The creature walked through a green field. It had left the ductwork some time ago. It had also grown considerably, feeding on a variety of insect pests living in the air ducts. In fact, it had been a struggle to get through the air duct screening. It had suffered some minor injuries forcing itself through, but it was healing rapidly.

The green it walked through was wrong, somehow. The colors felt "right," but the magnetic fields did not match anything it expected to find. It wasn't sure what kinds of prey would reside here, what kinds of danger it might confront. The colors slowly washing over its flanks reflected anxiety.

It could sense motion; there were things here too big to be prey. There were insects, too, but they were no longer of interest. They were too small, and would not yield enough sustenance to justify the effort to hunt them. It needed something bigger, and it needed something soon. The others would be growing, too. It needed to keep pace with the others, or it risked becoming prey.

The lateral sensory pits picked up movement. It slowed, and it raised its sensory cluster as high as it could, aiming its optics in the direction from which it felt motion. It began to feel an electromagnetic field. There was something familiar about that field. It was reminded, vaguely, of the host, the original prey from which it had hatched. That was good; *that* prey was delicious. That prey generated so many good sensations, so much fear and pain.

It could see movement. The prey was gray. It periodically paused and raised a pink tip into the air. It pulsed, suggesting it was sampling the air. This prey's surface was different; it was covered by countless, thin, little threads. The host had some threads like that. It didn't like the threads; but, at least on

167

the old host there was delicious food under the threads. And this prey didn't appear to have a shell like the other food encountered in the air ducts.

The new prey was coming this way. The creature stopped pulsing colors and matched the color and texture surrounding it. The green color was easy, but the texture was challenging; the green was long, thin spears of tissue. It did its best, forming ridges that resembled the edges of the spears.

The prey was coming this way, but would pass by at some distance. It slowly, ever so slowly moved to the side, pressing through the green, positioning itself in the path of the prey. It slowly retracted its sensory cluster, holding it close to its body. It could hear the movement of the prey, approaching through the green. It picked up scent particles. The vibrations were very exciting. The electromagnetic pulses were definitely like the original prey. It began emitting electromagnetic pulses as the prey approached, hoping to confuse the prey as it moved slowly into its path.

The prey was close enough for examination. It had four limbs, thicker than the prey from the air duct. They were covered with the thin gray threads except for the ends. The ends had grasping structures, and came to sharp points. The creature would need to be careful of those. The prey had a long, thin tail without any of the thin gray threads. The front, the part with the pink structure, also had two fanned structures and two moist structures that were round and black. The predator wondered if they were sensors.

Its claws ached to strike; its body trembled from excitement, and the strain of maintaining camouflage. But it waited, silent and motionless, and sent electromagnetic pulses to confuse the prey. The prey came closer. *Closer...*

Its claws shot out and grasped the prey behind the wet, pink structure and right about its midsection. The prey screamed, making a shrill sound that excited the hunter. The prey was so strong! Its legs were so strong! The prey tried to jump and run. It could feel the prey's mouth, clamped by the left claw, struggling to open. It could see the sharp, white structures in the mouth. *Oh no.* It had not noticed the prey's mouth before; it feared what those sharp, white structures could do if the prey got loose.

But the prey couldn't get loose. Its claws held fast, and reported valuable information. There was no shell; the body was warm, and it could feel muscles moving under the skin. It could feel something pumping inside. It could see and feel the body expanding and contracting as the prey brought air into itself and expelled it. It could hear the prey making shrill sounds. It could feel the prey's fear; it could sense how much pain the prey was experiencing. The prey's struggles began to wane. It was tiring. It could feel, with its right claw, that warm liquid was coming from near the base of the prey's tail. The liquid was acidic, an ammonia-based product. It was releasing its waste products as it struggled.

It began to pull the prey towards itself, enjoying its resistance, feeling the feet dig into the soil beneath the green blades to no avail. It was too heavy for the prey, and the prey could not resist the strength of its claws. Bright reds and yellows and blacks pulsed rapidly along its flanks. It stopped sending dampening electromagnetic pulses; it wanted the prey's mind to clear. It started sending pulses that would intensify the prey's sensations.

It wanted the prey to experience *everything* that was about to happen.

As the prey neared, it realized the round things near the pink tip and fanlike structures were optical sensors. They were eyes. It moved its sensory cluster and aimed its six eyes into the prey's eyes.

It sensed panic.

The predator drew the prey in close, then opened its beak and bit deeply into the prey's flank. It did not stop staring into its prey's eye. It didn't like the sensation of the thin gray threads, but it could feel skin and muscle tearing as it bit. Red fluid that tasted like iron flowed around its beak, and the prey shrieked loudly.

So exciting.

It bit and it tore. It reopened its beak and launched its tongues against the flesh of the prey, tearing off bits of the prey. It stared into the prey's eye. It could feel the prey's pump beating rapidly; it could feel the muscles of the legs struggling. It could sense the extreme pain and the panic; the screams and its struggling were absolutely delicious. It was *so* much better than the shelled, crunchy things it had been hunting in the air ducts.

It did its best to keep the rat alive as long as it could, staring into its eye, reveling in the experience. The creature was almost delirious with delight, loving every moment as it slowly, steadily consumed it. Its beak and tongues dove deep into the rat's flank, carefully rasping flesh from inside the rat while striving to avoid anything essential to the rat's viability. After a time, it could feel the pump's beating becoming irregular; it could feel the electromagnetic pulses weakening. At long last, as the rat lost its grip on life, it began to methodically dismember the prey. Finally, the creature everted its digestive organ to harvest the last bits of sustenance. It did its best to lap up the body fluids before they soaked any further into the soil. It discarded the larger bits of bone as it finished cleaning them, leaving a scattered heap of bones and gray hairs on the ground.

It stretched its sensory cluster again, and continued on its way. The hunter anxiously desired more prey. It was growing, and it was *hungry*.

▪ 16:17

Sylvia entered the hangar deck wearing her environmental suit undergarment and found Nikko overseeing preparations. It was the first time they would be deploying the submarine.

It was, essentially, a four-meter-wide transparent sphere sitting on a pair of bright yellow torpedo-shaped propulsion units; an array of sensors and robotic arms sprouted from the forward underside of the sphere. A rear ladder led to a wide, bright blue platform surrounding the entry hatch to facilitate entry and exit. The underside of the platform was rigged with a variety of lights and more sensors.

Nikko greeted her with a kind smile and a quick wave while supervising preparations for loading the submarine into a landing craft. A stout arm telescoped with a dangling hook out of the landing craft's furthest cargo bay door. Two of Nikko's operations team stood on the submarine's platform; when the arm reached them, they guided the hook through a loop integrated into the sphere behind the hatch.

Under their guidance, the slack in the cable tightened. She could hear a slight groan and saw the rear of the landing craft dip ever so slightly as the arm took the weight of the submarine, slowly lifting it from the loading area. Once it had a meter's clearance, the arm retracted, bringing the submarine into the landing craft.

Once the submarine set down in the landing craft's cargo bay, Nikko visibly relaxed and turned to Sylvia while the operations team secured the submarine. He immediately launched into details without prompting:

"Normally, we move heavy objects with anti-gravity suspensors, but we need to be able to deploy and retrieve the submarine at Eden's ocean surface, so we have to use the arm." He laughed, then added, "it was *so* much easier to move it with a shuttle tug between the *Lucky Strike* and the colonial frame! In the zero-G vehicle storage aboard the ship…" He extended his arms to demonstrate, "you just gently grab the thing with waldo arms and carry it to the zero-G vehicle storage on the colonial frame. *But,* when you want to bring it in here to prep it, you have to load it into the shuttle tug cargo hold and *then* deploy it with a cargo arm like you just saw. After that you prep it, then load it into the landing craft." He laughed again, then sighed happily. "Sometimes zero-G is just so much easier!"

Sylvia laughed lightly, "well, we're grateful!" She looked around the hangar deck and asked, "am I the only one on time?"

Nikko checked his clock, "actually, you're a few minutes early. The others should be here any time." He looked at her undergarment. "I gather you're planning to wear your environmental suit?"

She shuffled uncomfortably at the attention to her attire, but managed to shake it off with a light laugh. "My understanding was that we'd either wear the suits on the landing craft, then board the sub after deploying; or we'd have to be in the sub for the duration of the expedition from leaving to returning to the hangar deck."

"Personal choice. It's a little more comfortable boarding at the surface, and the environmental suits aren't bad inside. And once you seal the sub, you could, in theory, open the visor." Nikko looked into her eyes and asked, "how are you holding up?"

She remained silent and looked away; after an icy minute, she returned his gaze. "Oh…you know. Or maybe you don't; I don't know, I hope you don't." She shrugged, and her eyes glistened with the threat of tears. "I haven't lost anyone before…a team member, I mean. It's awful, losing a friend, but it's even worse when you're in charge. Everyone's safety is *your* responsibility. You know?"

Nikko put a hand on her shoulder. "Yes, I do know. I've been doing this a long time. People get hurt in this line of work; people have to assume risks to get the job done, to keep the ship safe. I've had to give people dangerous assignments, and sometimes their luck runs out. That's different; you all faced risk. You didn't ask him to do something you weren't willing to do yourself. You didn't put him in any more danger than you put yourself. He knew the risks, and he loved his work. Honor that."

She looked gratefully into his eyes, then lowered hers and nodded. "Thank you."

He looked past her, then said, "here's your group."

She turned to see Bharat, Gwen, and Enrique. An ecologist, a microbiologist, and a seismologist: an appropriate mix for their first expedition. After they exchanged greetings, Enrique said, "I…we just heard. I'm sorry for your loss."

Bharat and Gwen bowed their heads briefly. Sylvia nodded, "thank you. He was a good man."

A young man approached that she didn't recognize, also dressed in an environmental suit undergarment. He greeted Nikko, then introduced himself as their submarine pilot, Harrison Chang. He then added that he preferred "Harry."

Nikko informed them that their environmental suits and a set of life support packs were already loaded aboard the landing craft. They walked up the cargo bay ramp, verified their suits were properly packed, then climbed to the upper deck and took their seats. Nikko and his operations team left the hangar deck. The operator and landing craft pilot went through the launch process and, as the landing craft left the hangar deck, the team endured brief weightlessness as the landing craft headed to Eden's surface.

Harry turned to Sylvia and asked, "you're the oceanographer, right? Where are we going?"

"Yes, I'm the oceanographer. We're heading to a trench, a subduction zone between two tectonic plates. We'll be looking for any tectonic activity, and we'll be looking for deep sea life forms."

Harry shook his head and laughed. "I'm just a pilot; I have no idea what you just said."

Enrique also laughed. "Okay, the planet's surface is a series of plates floating over the molten core. The movement of the plates is driven by the gravitational tugs of the moons." He held up his hands, palms down, and moved them back and forth to demonstrate edges rubbing against each other. "As the plates move, they rub at the edges and that causes quakes and volcanoes, and, sometimes, when one slides under the other, it lifts to form mountains or drops to form a trench." He turned his hands so that the edge of his right hand slid under the edge of his left, then said, "on Earth, such trenches often had volcanic vents. Such vents release energy that fuels ecosystems where sunlight can't reach. We want to know if Eden has anything equivalent."

Sylvia continued after Enrique finished. "We've been running marine drones across Eden's oceans, using sonar to map the ocean floor. We've mapped about twelve percent of Eden's oceans' floor at this point, and we've found our first trench. We can't know how many trenches there are or whether there are deeper ones, but this one is over 9,000 meters deep. That's certainly deep enough to warrant exploration."

Harry nodded. He knew they were expecting to go to 9,000 meters depth. That would be an astounding, crushing 900 atmospheres of pressure, or about 930kgf pressing against every square centimeter of the sphere. It would take four hours to descend, followed by three hours exploring the seafloor, then another four hours to ascend. It will be a long day: a fascinating day, but a long one.

As the landing craft descended and the tug of gravity resumed, they made small talk, debating the relative merits of their preferred colonial pubs and dining establishments, how they compared to what they'd left behind, current entertainment available from other colonies via subspace communications, and so forth. They made their final descent to the ocean surface near the equator in the southern hemisphere, at least a thousand kilometers from the nearest land mass. It would not only be their first expedition to an oceanic trench, it would also be their first exploration of the open ocean.

It was late afternoon when the landing craft hovered a meter above the ocean surface. The first moons had risen above the eastern horizon.

Harry spoke with the landing craft pilot, then donned his environmental suit. He went through the airlock at the rear of the upper level, and descended to the cargo bay. The science team began donning their environmental suits to follow.

The landing craft pilot lowered the cargo bay door and Harry deployed a drone that would circle their descending submarine and alert them to any large approaching life forms. He then untethered the submarine from the cargo bay floor, then went to the control panel to lift the submarine. The arm extended out over the water, then gently lowered the submarine, which began to gently rock with the light waves. The arm retracted slightly, pulling the submarine closer until the platform was near enough to the lowered cargo bay door to allow the explorers to walk out of the cargo bay and step across to the platform.

Harry went first, opened the hatch, then pointed at each of them in turn, so they could descend through the hatch to their seats. Gwen and Bharat sat in the rear two seats, Enrique and Sylvia in the middle two seats, and Harry, after he released the arm hook and closed the hatch, took his seat in the front.

The seating was raised in the back and lowered in the front to give each occupant a clear view through the walls of the sphere. As he took his seat and began manipulating the controls, Harry said, "I know the rocking is a bit disquieting, but once we begin descending, it'll be steady."

Gwen laughed, her face turning a little green. "That's a relief!"

Harry nodded, and began to pilot their submarine downward as their landing craft began its ascent. They watched the water line rise around the sphere. A cloud of small, shiny creatures swam past them, followed a few moments later by a larger creature, presumably hunting the smaller ones. Bharat laughed with excitement. "Beautiful! I've no idea what they were though…are we recording?"

Harry sighed, flipped a switch, and said, "we are now."

Bharat laughed again and thanked him.

They descended steadily. Sylvia looked upwards and could see, through the undulating surface, a distorted disk with rays of sunlight reaching down towards them. She looked down between her feet and could see nothing but greenish gloom. She looked to the side and saw tiny bits of organic matter, some moving, some not, reflecting the receding sunlight as they floated past.

"Not nearly as much sea life as we saw nearer the shore," she commented.

"That's to be expected if Eden is anything like Earth," Bharat responded. "Most of the ocean's life is near the shorelines."

They continued to descend, and the light became dimmer and dimmer.

"One hundred meters." Harry calmly reported their depth, and continued to monitor the submarine's status. "We'll need to turn on our lights shortly."

After another ten minutes of descent, the lights suspended from the platform snapped on, illuminating their surroundings. It startled some creatures in the distance that swam away before they could be seen.

"200 meters."

A beeping sound caught their attention. Harry looked to his left and then pointed to the screen, saying, "The drone spotted something large heading our way. The lights may have attracted its attention. There's no light at this depth."

Bharat touched a control and a projection of the approaching creature appeared before them. It was roughly spherical, with long tendrils floating outward in all directions. There were no perceptible sensory organs or, for that matter, a clear front, back, top, or bottom.

"Can we move towards it?" Bharat asked. Sylvia tried to ignore the pit suddenly growing in her stomach.

Harry, looking slightly to the left, said, "sure. But it will slow our rate of descent."

As he worked their steering and dive plane controls, they could feel the submarine move in the direction they were looking. It was somewhat

174

disorienting to feel that motion without any associated visual perception of motion; the complete lack of discernible reference points made some of them a little queasy, especially Gwen.

After a few moments, Gwen pointed and asked, "is that it?"

As they progressed, the object grew. Enrique asked, "how far away is it? How big is it?"

Harry looked at the submarine's screens; tiny lines of data began to flow in front of his face. "The drone indicates the main body is some ten meters in diameter. The tendrils vary in length, up to fifty meters." He turned to Sylvia and exchanged looks. "I'm not sure how close we should get; I'm not sure what happens if we touch a tendril."

As she looked out into the gloom, her face became pale and rigid. "The submarine is impenetrable, right?"

"According to the specifications," Harry answered with a grim smile. They laughed nervously, except for Sylvia.

They watched in wonder as they approached, the size of the thing becoming more apparent. Its skin was transparent; its interior a maelstrom of moving colors.

As they observed, a small creature, possibly dazzled by the lights, touched a tendril. The creature immediately began to flail and dart around with distress, but was held fast by the tendril. Coiling into a spiral, the tendril drew the creature closer to the body. Eventually, as the tendril completely flattened and coiled flush with the body surface, an opening appeared in the body and the struggling prey was pinched into a small bubble. Trapped and crushed, the tendril pushed the creature straight into the pulsating maelstrom.

As they watched, the small creature stopped moving, then began to dissolve. They realized that there were other such bubbles, of varying sizes, moving through the interior of the creature.

Bharat shook his head, "fascinating!"

"We really need to resume our descent." Determined to stay on track, Harry took to the controls as his passengers watched the creature disappear above them.

They continued to descend into blackness, their lights illuminating the region surrounding them. Occasionally something small that would dart away, leading to speculation and discussion. The drone would report life as well, sending images that they projected and contemplated. There was a broad diversity of forms.

As they passed 7,400 meters, they briefly encountered a creature about twenty centimeters in length, bearing a light apparently derived from one of the eyestalks on its sensory cluster. It also had a large beak, presumably equipped with tongues that would help it capture and hold prey. Bharat and Sylvia compared the creature with notes on the specimens collected at the tsunami site. Eventually, as they descended past 9,000 meters, they began to see a bottom beneath them. Harry reduced their speed, and for the first time, their physical sense of motion was matched by a visual sense as they saw rock formations passing by.

The formations were irregular but low, interrupting otherwise feature-less spans of sand. They followed a gentle downslope. Periodically, something would swim beneath them, nearly always pale in color. Some were clearly of the trilaterally symmetrical super-kingdom, but they lacked, or at least did not bother to deploy, the eyed sensory clusters common to their brethren on the surface.

They began to see more signs of life. Motion in the sand suggested bur-rowing bottom dwellers; slow swimmers nuzzled the bottom, presumably rooting out some poor creature for a meal.

Harry noted, "the outside temperature is climbing."

Enrique checked his screens, and confirmed Harry's readings. "We must be approaching the vents. Be careful, the water near the vents will likely be superheated."

"I'll approach carefully, but I won't expose us to anything above 50°C."

The rock formations were increasingly encrusted with sea life, strange fanned and conical structures. Gwen observed, "the water must be carrying more microorganisms; the fans and cones are most likely filtering them from the water for food. I'm taking some samples."

They continued and the structures became larger and more elaborate. Finally, they could see spots of orange light in the distance. Enrique pointed and, with a slightly raised voice, said, "there! That's the vents. You can see volcanic activity."

Harry nodded, "yup, and this is about as close as we'll get."

Something caught Sylvia's eye. "Can you see that shimmering quality in the water? That's likely where the water is warming sharply."

"And that's what's probably feeding this ecosystem!" Gwen tapped furi-ously on her computer screen, making notes and taking samples simulta-neously. "There must be microorganisms using heat or chemical energy from the vents and serving as the foundation of the food chain down here. Very similar to the deep oceanic trenches of Earth!"

They continued to explore the area, capturing images, and using the waldo arms to take samples.

Eventually, Harry cut them off. "I'm sorry, we need to begin our ascent."

They began to rise, leaving the teaming ecosystem behind them. They climbed through the relatively lifeless void, discussing and debating what they'd seen.

Much later, as they neared 1,500 meters, the drone alerted them that something large was approaching. They brought up a projection and saw a long, torpedo-shaped creature with bilateral symmetry and a large, vertically oriented mouth bearing rows of sharp teeth.

"How big is it?" Enrique asked in alarm.

Harry pushed a button and an image of the submarine appeared next to it, dwarfed by the creature. He whistled, "not much we can do about it; we need to ascend. We're secure in this sphere, but I'd feel better if we all strapped in just in case."

They peered into the gloom, then began to perceive a shape moving slowly towards them. As they stared, it swam past them, apparently unaware of their presence. Its eye was nearly as big as the sphere, and they could feel their vessel move in the wake of the creature.

Bharat's voice expressed his awe. "That is, by far, the largest creature we've encountered on Eden so far!"

"Yes, and I'd like to not encounter it any further," Harry spoke through gritted teeth. He paused, then switched off the lights. "I don't know whether the lights were attracting it, but let's not take any chances." They continued to ascend.

Another alarm went off. The drone reported that the creature was returning, *fast*.

They stared into the gloom, then realized it was heading straight for them. As it accelerated, they saw the jaws spread open wide and clamp against the side of the sphere. The sudden force pushed them sharply backwards as they screamed in shock and surprise.

The sphere shook side to side. Then, suddenly, the creature released them from its jaws and circled around. After a pause, it swam calmly by, staring into the sphere as its eye passed. The whole crew watched with extreme relief as it disappeared back into the deep.

After a long pause, Gwen said simply, "wow."

"Yeah," Harry agreed absentmindedly as he tended to ringing alarms and system readouts. Finally, he breathed a sigh of relief. "We're good. I thought we would be…we survived 900 atmospheres, so we should be okay against that thing's mouth. But, still…"

He asked whether anyone was hurt; they had a few bruises from being knocked about, but nothing noteworthy. They resumed their ascent. After the drone reported that the creature had moved on, they reactivated the lights.

They continued to see smaller sea life around the sphere. Suddenly Bharat spoke up, "I can't imagine what that thing feeds on out here."

"What do you mean?" Sylvia asked.

Bharat cocked his head thoughtfully and looked out of the submarine. "Well, as we discussed earlier, on Earth the most productive marine ecosystems are along the shorelines. Something like that needs a lot of food, and, as you can see looking around us, there isn't much out here."

"Perhaps it doesn't normally spend time this far out," Gwen contemplated, "or perhaps it crosses between shoreline ecosystems. Or, maybe there are ample microorganisms near the surface to support an ecosystem of filter feeders. On Earth, before they went extinct, whales were immense creatures that fed on tiny creatures near the surface."

Bharat shrugged and said, "we'll figure it out over time."

Finally, they breached the surface and summoned the landing craft.

Harry climbed out to connect the hook, then helped each of them step across to the cargo bay door. They groaned and stretched after so many hours in the sphere. Once the sphere was loaded and after they'd lashed it down, they inspected it for signs of their encounter. The sphere was covered in scratches; deep dents ran along the edge of the platform.

It was Enrique that found the day's treasure: a massive tooth lodged in one of the propulsion units!

•DAY 22, 05:17 PLATFORM TIME

As she woke up, Jian realized that she did not feel right.

She was lethargic, which was strange, considering she had gotten a good night's sleep. She should be more energetic. She'd even started exercising lately, and she liked the results. She was thinner, or rather leaner, with more muscle definition.

She was enjoying the results, as was her favorite. She hadn't seen him in a few days; she needed to call him, see if she could entice him to another dinner. She really enjoyed him; he was young and enthusiastic, and she loved his smile.

But she just wasn't feeling right. Something was off.

Well, it was probably nothing…but why take chances?

She threw her legs over the edge of her bunk, pulled on her shoes, and stood up. *Whoah!* She felt *faint*. She really, really wanted to just lie down and go back to sleep. Which she decided she would do, just as soon as she returned from Sick Bay.

• 06:38

Elke was relaxing in her bunk, starting her day slowly, when the call came in.

Daniella Wu simply said, "Elke, we need you. Please come to Ring 6, Section C, Room 22."

She sat up and responded, "what is it?"

"Best we discuss in person. I'll be waiting."

"Fine, on my way."

She sat up, sighed, and pulled her clothes on as she stood. *What could be so urgent?* She sat back down to don her boots, then stood again and fastened her collar. She walked into the hallway and then to the nearest lift. She was quartered in the command rings and took the lift to Ring 6, one of the levels dedicated to the living quarters and workspaces of the science teams. She turned left, exchanged greetings with passersby, and ducked through section bulkhead doorways. Eventually, she reached Dani and found her standing by a door with John, David, and two men she presumed to be from Dani's security team. John looked a bit shaken; David was a wreck.

She folded her arms. "Okay, what's going on?"

Dani, expressionless and monotone, replied, "we received a call from one of the geology lab technicians, Natalia Dileep. She said she was concerned about one of the planetologists, Esteban Kay. Apparently, they'd spent an evening together a few days ago, then he sent her an invitation to dinner but never followed up with specifics. She tried to reach him periodically over the next few days, but he never responded. Earlier today, she contacted his supervisor, David here, and asked whether anyone had heard from him. David had not but was not concerned because Esteban usually worked independently. In any case, she then came here to his private quarters and tried to reach him. When he did not respond, she became concerned and called us."

Dani looked down and sighed, then continued, "I came down and overrode Esteban's privacy, then we entered." She paused, then continued, "Natalia became hysterical, so I called John."

John said, "I brought one of my nurses, and he took Natalia to Sick Bay and sedated her."

"Right, and then I called David, and now I've called you." Dani paused, then, looking into Elke's eyes, added, "brace yourself, and step carefully." She turned to David and added, "David, I'm going to suggest you wait out here." He nodded, head bent, eyes closed, and arms folded.

She touched the door controls, and the door became translucent. They stepped through.

Dani stepped carefully, and Elke followed suit, matching her footsteps. The floor was mostly clear, but there was some debris and they did not want to disturb anything. Elke followed her into the bunk area, then stared.

John came into the room behind her, and whispered, "note that the medium and larger bones are mostly still in place, but the smaller bones are not. Some are cracked; some are gone. We think we'll find them as we go through the room. Note that there's absolutely no tissue left on the bones; they're perfectly clean. No tendons, no ligaments, *nothing*. The long bones have places where they've been notched somehow, and the marrow seems to be..." John stopped, noticing the look on Elke's face.

Elke stared at the skull, propped up on the pillow, empty eye sockets aimed downward, looking over the body. The teeth were in place, as were the cervical vertebra. She stared at the pelvic region; his briefs were still in place, incompletely covering his pelvis. It took her a moment to recognize what was on the pillow around his skull. It was his hair, pretty much where it would have fallen around his head if his flesh had just suddenly vanished.

"It's...the damnedest thing I've ever seen." John still wasn't certain if she was listening, but he pressed on. "You can see blood stains on the sheets, but not that much really. Not what you'd expect from an adult body. There are a few, very, very few bits of tissue here and there on the bunk and around the room. And, I'd like to call the biologists in here because, and I can't be completely sure, but I think some of the other materials are droppings."

Elke paused, then looked at him incredulously. "Droppings?"

John nodded, "yes, droppings."

Elke paused again. Her ears were ringing, and a cold sweat started to set in. She looked up and said, "George? George Sied, are you on board?"

After a moment, his voice could be heard over the room speakers. "Yes, Governor. In my quarters. What can I do for you?"

Elke, looking down at the body, simply said, "please join me immediately. I'm in...wait, Dani, where are we?"

"We're in Ring 6, Section C, Room 22."

There was a pause, then George responded, "on my way."

There was an uncomfortable pause in the room. No one knew where to stand; no one knew what to say. There were many side glances at the skeleton in the bed, as they all tried to process what could have happened there. Finally, Dani spoke up:

"Normally, I'd be thinking in terms of a crime scene; but I'm not thinking that here, so I'm not worried about forensic evidence. Still, I think it's best we disturb as little as possible until George arrives. Images have already been captured, of course."

Elke nodded slowly, staring at the bones. She then turned to Dani and asked, "assuming, for a moment, that the 'droppings' suggest a predator, have you isolated this ring segment?"

181

"As you saw on your way here, the bulkhead doors between segments are still open. I wouldn't expect anything to be able to walk through the halls without getting noticed. Each air vent between segments is screened. The screens are fairly fine; anything big enough to attack a human wouldn't be able to get through them. Closing them is possible, but keeping them closed is problematic. Ring segments don't have their own life support systems. On the other hand, the air ducts for this room still have their screens in place, so if our mystery guest can't get through our screens, then where is it? And to anticipate your next question, I ordered a well check for the other quarters in the segment. Everyone currently in their room is fine; occupants of empty rooms have been located and they are also fine. Their rooms are currently being inspected."

"I appreciate your actions and anticipation, in light of everything, but I'm thinking we still need to enforce isolation. Without compromising life support, what can we do?"

"I checked with Moira. Every ring's ducts are connected with core life support in four places. I can block the ducts between rings and between the four quarters of each ring. The colonists won't notice anything other than some minor pressure gradients when they pass through bulkhead doors."

"Do it." Elke's face became grim and serious. "We'll reopen as quickly as we can. And please give some thought to how we might impose further isolation. I agree with you on the bulkhead doors, but I also know traffic levels vary. Is there something we can do to monitor who, and what, is moving through those doors?

Dani nodded, "absolutely." She paused, looked up, and said, "close all interconnect ducts between rings and ring quarters, and institute continuous monitoring of all bulkhead doors." After a moment, a computerized voice responded, "tasks completed."

Elke paused, then asked, "is there any possibility something's escaped from one of the bio labs on the hangar decks?"

John looked at Elke with extreme alarm. Dani thought for a moment, then responded. "Extremely unlikely. We maintain very strict quarantine discipline. There are air locks between the hangar decks and the ring supporting them, as well as between each bio lab and the rest of its hangar deck. Each hangar deck's life support is ducted separately, or we wouldn't be able to manage pressurization and depressurization. Finally, if something *had* escaped...well, the hangar decks are in Ring 40, and this is Ring 6. That's a long distance between the labs and our victim."

As she was finishing, George entered the room. "I wasn't far; my quarters are a couple of..." George froze mid-step and rapidly scanned the room around him. His face turned ghostly white, and for a moment, Elke thought he was going to faint. He stared wide-eyed at the bones, then slowly rotated his head to face Elke.

"What...the hell...happened *here??*"

Dani and John relayed what they'd told Elke as George listened silently, staring at the bones. After they finished, there were a few moments of awkward silence until Elke said, "George?" He turned to her. She caught his eye and asked, "can you tell us anything? Have you seen any creatures on Eden that could do this? Anything in any of the labs?"

George shook his head. "Remember, everything on Eden is biochemically incompatible with us. Nothing on Eden should have been able to consume Esteban. Beyond that, how would anything from the planet get into this room?"

"So we thought, yet here we are. I agree with Dani that this is unlikely to be the work of anyone in the colony. I've seen murders over the years; they're rare in the colonies or aboard ships, but they happen. But I don't see how anyone could reduce a human body to...this. John?"

John stood over the bed, shaking his head. "No. These bones are clean and perfectly placed. Cleaning all the tissue from these bones would be a challenging task, and the placement would require an advanced understanding of anatomy. Plus, there are the notches on the long bones and the extracted marrow."

"Understood, but..." George looked at him and gestured at the remains. "What's your alternative? The typical vermin that make their way onto deep space vessels can't account for this!"

"Enough." Elke turned away from the bed and bones; her jaw was set, her eyes brimming with pain and remorse. "George and John, I'm asking you two to work together. Take and analyze samples, and figure out what happened here." She looked them each in the eye and added, "we need to know, and we need to know fast. I want a briefing in four hours."

George and John looked at each other. John said sheepishly, "I can't promise we'll have it all sorted out in four hours, but we'll report everything we can at that time."

"Understood. Get to work. If you need anything, anything at all, let me know." She turned to Dani and said, "please support them in any way you can. I would also appreciate it if you would review Esteban's activities since we reached Eden. I want to know everything he did and who he did it with. And I'd like your recommendations as to how we keep this from happening to anyone else."

Dani nodded, "absolutely."

"We keep this to senior staff *only*," Elke added. She paused, then touched her wrist computer and said, "convene senior staff meeting in my conference room in four hours."

She turned to John, "I'm going to get out of everyone's way. I'll be in my quarters. Please have your staff look after David, and let's make sure neither he nor, what was her name? Natalie? Natalia? Yes, Natalia. Let's make sure they don't start a panic. Keep them both in Sick Bay for now."

John nodded and Elke, after a pause, walked out.

▪ 11:00

Roughly four hours later, she stepped through the portal to her conference room. Moira, Dani, Nikko, John, George, Cecilia, and Jenna were already there. The mood was somber, and fearful.

As she sat down, she asked, "how is David?"

"He's taking it hard," Cecilia responded. "He worked with Esteban for years. John's team is taking good care of him. I told him I'd cover this meeting, but I expect he'll be back on duty soon."

Elke nodded. "Very good. Well," she said, looking round the room, "I am assuming everyone has heard, by now, what we found earlier today. John and George, please proceed."

John stood, "I'll go first." He tapped his wrist computer, and Esteban's remains appeared on the table before them. Cecilia and Jenna grimaced and looked away. Moira tightened her jaw, but Nikko turned green like he might be sick. "Forgive me," John asked sympathetically, "I know this is unpleasant."

He sighed, then continued, pointing at the bones as he spoke. "Note that the remains are, with few exceptions, lying as if Esteban died suddenly and without violence. Also note the limited blood stains on the sheets and his briefs; it's not nearly enough to account for a body's normal quantity of blood. There are no signs of struggle, thrashing, anything. It is as if he'd simply laid down with his head propped up, and then everything but his bones and hair disappeared.

184

There are a few exceptions. Some of the phalanges, the bones of his fingers and toes, have been moved." He highlighted these bones, most of them still laying on the bunk near the hands or feet from which they had been separated. A few appeared elsewhere in the room. "In addition to the phalanges, some of the metacarpals and metatarsals were similarly disturbed, as was one patella, or kneecap." They were now highlighted as well. "All of these bones are clean; there is no residual tissue on any of these bones. If this were done by a human, it would require dipping the bones in a strong, lab-grade solvent, and then putting them back. To do so with such precision would be so challenging for anyone, even with a thorough and detailed knowledge of human anatomy that I would disregard this scenario as extremely unlikely.

The bones that contain marrow show some damage; at irregular intervals along these bones, there are small holes that appear to have been made by something sharp applied with substantial pressure. The marrow appears to have been extracted through these holes."

He raised his hands and said, "that is...really about all I can tell you. I can't definitively determine a cause of death, other than to attribute his death to whatever process removed the tissues."

He tapped his wrist computer and, as he took his seat, the display dissolved.

George stood and tapped his wrist computer. The display reemerged as an enhanced image of a pile of droppings. He said, "When we were in the room, John suggested these might be droppings. He is correct. Note also that this dropping includes this white material; it's similar, though not identical, to the uric acid produced by some Earth animals."

Jenna interrupted, asking, "what is uric acid?"

Jenna's expression revealed her distaste as George explained, "it's basically a dry version of urine. We all produce nitrogenous wastes. Aquatic creatures just dump ammonia. Animals like ourselves spend a little metabolic energy to produce urea, which costs us some water as we urinate. Some animals, like birds, spend more energy to save more water by producing uric acid."

He pointed at the white material. "As I said, it's not exactly the same as the uric acid produced by Earth organisms, but it's very close. Normal deep space vermin include some rodents, which do not produce uric acid, and arthropods like insects, which do produce uric acid but do not deposit it in this manner. I did a quick check, and the only organisms I can account for that produce and pass uric acid in this manner are the farmers' chickens and some of the colonists' pets, including a few small birds like parakeets and a few small lizards and snakes. None of them could account for this, and they would have produced Earthly uric acid, not this material.

So, we took samples of the droppings and of some of the tissues we found in the bunk and around the room." He paused, looking very uncomfortable. "We were able to confirm the remains were of Esteban Kay, and that the droppings did contain residual DNA from Esteban. Whatever creature we are dealing with, it did consume, and digest, Esteban's tissues."

He flicked his wrist computer and the image vanished.

"Which…is…impossible." George said this so softly, it was unclear whether he said this to everyone, or to reassure himself. He collected himself and continued in his normal voice. "We are *entirely, biochemically* incompatible with Eden's life forms. Even if something from Eden found itself in Esteban's room, and if it chose to attack and kill Esteban, any tissues it consumed *should* be passed undigested."

George dropped into his seat as everyone weighed the implications of what he was telling them. "While it remains impossible for anything on Eden to consume Esteban's tissues, we must confront the reality that whatever produced these droppings did, indeed, digest some of his tissues. And…we have detected the presence of non-Earthly tissues in the droppings as well. The digested material also included genetic material belonging to one of the five organochemical families we've found on Eden."

He flicked his wrist computer again, and the image of a petri dish with teased material rotated before them. A portion of the material became highlighted and the resolution increased. Inside the dropping materials were small shards of stiff tissue. "These tissue shards are also not part of Esteban; they are clearly from Eden. Biochemically, they are from the same organochemical family as the digested material we found."

He paused, but there were no questions. No one blinked, or spoke, for a while; a circle of stunned faces sat around the table.

Finally, Dani stood up. "You asked for an account of Esteban's activities since we arrived, so I reviewed the security logs. Esteban had been to the planet's surface twice. He was part of the first expedition to the tsunami site. He and a few other planetologists hiked above the tsunami inundation zone, found a promising placer deposit, and followed a stream for some distance looking for source rock, but didn't find anything of interest. Eight days later, he was part of an expedition exploring a temperate forest. They also found a promising placer deposit and followed a stream looking for source rock. This time they were successful; I'm told the deposit is likely to be rich. After the second expedition, he began working with samples collected by himself and other planetologists, working with the lab to prepare and analyze the samples. He also worked with the modeling teams to prioritize prospects and with the mining engineers to prepare proposals for drilling. All of his working time was spent in the planetological lab space on Ring 6."

Cecilia interrupted, "I'm sorry, do you maintain such records for *all* colonists?"

Dani responded, "as a security measure aboard deep space vessels and colonial platforms, our information systems monitor all personnel and all spaces. That information is collected and curated by our security systems and remains absolutely inaccessible until and unless a sufficiently urgent need arises. In order to access that information about any person or any space, a formal query must be logged by the captain or governor, the chief security officer, and the colonial leader. Elke, Jenna, and I logged such a request after our last meeting. We now have full access to Esteban's activities."

"Does that mean we can see what happened *inside* Esteban's quarters?"

"Unfortunately, no. The systems do record all activity, but they derive from that recorded material what information that might be pertinent, and, to protect our privacy, discard the rest. We know where he was; we know who he was with. If the system recorded, say, aggressive or violent behavior, it would have retained a recording of the incident. But whatever happened in Esteban's quarters, it was not perceived as "notable" by the recording systems and it was not kept."

As to his personal time, he had breakfast in his quarters and took his midday and evening meals in the Ring 6 commissary. He did not leave Ring 6, nor did he patronize any of the bars, restaurants, or other businesses in the commercial rings. He used the Ring 6 gym facilities on alternate days for the first week, not at all during his final week."

Dani cleared her throat. "Each of Esteban's evenings were spent in his quarters with one of five partners, including Natalia Dileep, the lab technician whose concerns led to the discovery of Esteban's remains. He has also spent some of his evenings with Jian Harris, with physiologist Corrina Niobe, with crewman Robert Esposito, and…" she cleared her throat again as her cheeks flushed bright red. "And for one evening, me."

She paused, then resumed. "I determined that we should interview these people, and that it would be best if I did not conduct the interviews. I tasked one of my female security team members to conduct the interviews with the women. We interviewed Natalia and Corinna. Jian said she is not feeling well, and we will be interviewing her later today. We were not able to arrange an interview with Robert Esposito. He is currently aboard the *Lucky Strike*. As of yesterday, their acceleration rate exceeded our ability to establish communications.

There are no indications that any of the women viewed their relationship with Esteban as anything but casual, or that these women were aware of each other. I know that, while I had no expectations of exclusivity with Esteban, I also had no specific knowledge of his other partners. He only saw Natalia once, but, per the invitation he sent her, he was planning to see her again. His most frequent partners were Jian and Corinna, seeing each one to two nights per week. His relationship with Jian started more than a year ago, and with Corinna about two months ago."

She sat down, then concluded, "I don't know whether this is of any significance, but, according to Corinna, Esteban had been losing weight. She reported he was getting progressively leaner, had visibly less body fat, but wasn't clear why. He speculated that it had something to do with physical exercise during his participation in the landing parties."

No one spoke for a few minutes.

Elke then said, "I realize you've only had four hours to work on this, and I appreciate the detailed analyses you were all able to conduct. They establish a solid ground for discussion."

She placed her hands on the table and leaned back, continuing, "George and John, please speculate. What do you *think* has happened?"

John looked at George, then said, "we think something from Eden, somehow, subdued and consumed Esteban in his quarters. It managed to accomplish this without sufficient violence as to trigger retention of the event by monitoring systems. It has, somehow, adapted biochemically to be able to digest his tissues, and is now somewhere aboard our colonial station. We need to prepare ourselves for another possible attack."

Elke nodded grimly, "I agree. I need you two to continue working on this. We need to understand what is happening in order to address it. George, I'd like some ideas on what this creature might be, what it might look like. I would ask that you pull in some of your team members, but I will defer to you as to which team members can make a contribution.

I want to convene our next meeting in another four hours. I know that's not much time, but we need to be able to take action on whatever you learn, so we must establish regular meeting intervals. Would every six hours be better?" She looked around the room, saw some nods, then said, "Six hours it is."

She turned to Dani, "please continue to keep the ring segments locked down. I would like more options for finding and containing whatever predator we've brought aboard. And we must determine *how* a predator was brought aboard in the first place!"

She then turned to Nikko, "I want all landing parties to return and further landing parties suspended until we know what happened. That includes suspending the plan to establish an encampment without environmental suits. I want all returning landing parties thoroughly inspected and quarantined. We'll need a reason…can anyone think of a good reason? Or should we reveal what's happened? I don't want to start a panic."

Jenna tilted her head, then said, "I'm sure we can come up with something innocent sounding. Maintenance issues? Eden's weather concerns?"

Elke leaned back, pursed her lips, and exhaled after a long pause. "No, I think we have to be candid. We can keep it high level, but if we aren't candid, we risk being contradicted by the rumor mill. We can't risk losing credibility. Candor's the lesser evil. Unless someone can convince me otherwise, we'll publish, on the daily news feed, that we've had an injury presumed to be the result of an escaped creature from Eden. That we're suspending landing parties until we understand how it escaped."

She paused, then continued, "Moira, we don't know what it is or how big it is. Can you fabricate weapons? I don't want projectile weapons; we can't have any stray shots compromising the integrity of the rings. How about something generating a burst of electricity?"

"I'll have to think about electricity," Moira paused pensively. "I believe we can produce a projectile weapon that might subdue a predator, but would not have the ability to pierce any station wall. Rubber bullets perhaps? I'll have options for you to consider at the next meeting."

"Very good." Elke stood, saying, "We all have a great deal of work to do. I very much appreciate everyone's commitment to the good of the colony. Report back in six hours."

As people left the conference room, Elke touched Dani's elbow. When they were alone, she put a hand on Dani's shoulder and asked, "are you okay?"

Dani looked down, nodded, then looked into Elke's eyes. "Elke, you know me. I'm a hard-core, deep space lifer; I don't want or need a committed relationship. I got to know Esteban on the outbound journey. I needed a sparring partner, and one thing led to another…" She shook her head and smiled, "I, well, I enjoyed his enthusiasm, shall we say. He was an attractive young man. But I didn't form any attachment to him, more than I have with the other young men I spend time with." She paused again. "It's definitely unsettling. It's rather strange to see the bones of a man you've had inside you. But, I'm okay, thank you. I'll be okay."

Elke nodded, "alright, good enough. But remember that I care, that there's no shame in needing a little time. If you decide you need to take a break, you let me know. And, if you need to talk about this, or *not* about this, you let me know. Anytime, okay?"

Dani nodded, but looked away from Elke's eyes. "Thank you, Elke. I'm good for now."

▪ 15:51

John and George stood side-by-side at a workbench in George's lab, investigating an array of samples taken from Esteban's room. John backed up and plopped into a chair, rubbing his face with his hands in exhaustion. George waved an arm across the samples and began:

"Okay, let's walk through this. Obviously, we've had very limited opportunity to observe Eden's wildlife in the field or in the lab. We've found five distinct organochemical suites amongst Eden's life forms. On Earth, there was only one. Every life form on Earth shares the same four nucleic acid base pairs, the same amino acids, the same stereoisomers of complex molecules. So, on Earth, we have the animal kingdom, the plant kingdom, the fungal kingdom, etc., all sharing the same fundamental biochemistry. Here, we're thinking in terms of five 'super-kingdoms' for the five organochemical suites. Each of these five 'super-kingdoms' have microorganisms, plant equivalents, and animal equivalents, and these 'super-kingdoms' have as much in common with each other as any of them do with Earth. Which is to say, nothing. Nothing at all. To the point that, based on our limited observations, it seemed that the members of each 'super-kingdom' ignore the other 'super-kingdoms' from a nutritional perspective. We've seen large herbivores walk by plants from other 'super-kingdoms' and only feed on plants from their own organochemical suite. Agreed?"

John, seated, started nodding. "Agreed. That's been our presumption, based on our limited observational experience. We need to be open to the possibility that this presumption has been self-reinforcing, that we've not recognized conflicting observations when we could not determine what 'super-kingdoms' each organism belonged to."

"Fair enough," George shrugged. "I might be willing to accept that, on a planet with five 'super-kingdoms', over time, a predator from one organo-chemical suite may have evolved the digestive enzymes necessary to be able to feed on members of another organochemical suite. But that doesn't explain Esteban! How would an Eden predator be equipped with the enzymes nec-essary to be able to digest tissues from an Earthly organochemical suite? I'm sorry, it doesn't make sense."

A voice announced, "John Schmitt, please report to Sick Bay."

John responded, "what is it?"

The voice continued, "you have a patient. Please report to Sick Bay."

John sighed, then turned to George. "I can't imagine. But duty calls. I'll check back after I see my patient."

George raised a hand and said, "I understand. I'll keep reviewing reports from the landing parties, see if anything makes sense. See you later."

- 15:53

John walked into Sick Bay, saw one of his nurses, and asked, "Aldo, what's going on that nobody else could handle?"

Aldo looked up and said, "you need to see this."

They walked into the first room on the left. Jian was lying on a medical bunk. A panel above her displayed a variety of diagnostic information.

John said, "hi, Jian, what brings you in?"

She smiled weakly. "I'm feeling pretty bad, John. I'm really weak and tired, and I don't know why. I noticed it this morning and decided I should come in to see what you all can find."

John nodded, "alright, we'll see what we can see. Anything else? Just lethargic? Any other symptoms? Anything else you've noticed the last few days?"

Jian sighed, "not really. I've been feeling fine, actually. I changed my exercise regimen about a week ago. The results have been good; I've gotten much leaner." She smiled, then added, "Surprisingly so, more than I would have expected."

John kept a calm face, but his eyes widened with alarm. He looked at the diagnostic displays; all of her vital signs seemed normal, except her metabolic rate. It was a little high. He looked back at Jian, "I'd like to scan you. You won't even know it's happening. Okay?"

Jian smiled, "sure."

John looked at Aldo, "take the scan."

"I already did." Aldo motioned to the door, and John read a genuine fear in his eyes. "Let's go review the results."

John put a hand on Jian's shoulder and said, "I'll be right back." She nodded confidently, but her face revealed her growing anxiety.

They walked out of the room as Aldo whispered, "I took the scan, reviewed the model, and called you immediately." They walked to John's office, and Aldo brought up an image of Jian on the desk projector. "Make the image translucent." The image became translucent. Then he said, "highlight foreign bodies in red."

A large number of thin, red, wormlike shapes appeared within Jian's body, moving slowly through her tissues.

John leaned forward and stared at them. "What the hell are *those*?"

"That's why I called you." Aldo and John locked eyes, and Aldo shook his head in disbelief. "I have absolutely no idea."

▪ 16:15

As John entered the conference room, Elke asked, "alright, John, we reconvened early. What have you found?"

John flicked his wrist computer, and the image of Jian in her medical bunk appeared over the table. "Jian Harris came to Sick Bay earlier today, complaining of lethargy. She said that she was feeling well overall, but was inexplicably tired. She also reported that she'd noticed she'd been losing body fat the past few days."

"That sounds like Esteban," Daniella gasped.

John nodded, "exactly, that's what I thought. We ran a scan. And, after checking a number of possibilities, one of my nurses suggested highlighting foreign bodies." He grimaced before tapping his wrist computer. "This is what we found."

The image of Jian became translucent, and the red worms appeared. Several gasps rang out across the table. Jenna turned her head in disgust.

George stared straight ahead, gripping the arms of his chair. Elke tapped the table gently, gathering everyone's attention. "Is Jian okay?" she asked.

John looked at her, "I don't know." He flicked his wrist computer again and the image of Jian was replaced with an expanded image of one of the worms. Now that it wasn't highlighted, it was brown instead of red.

"I moved Jian to an isolation unit," he continued, "and surgically removed as many of these worms as I could find. I can't say whether I've gotten them all. We will continue to observe her and, if our scans find more of them, I'll operate again. I'm honestly hoping we don't have to. These things were deeply embedded in her body, in her muscles and a number of her organs. The surgery was very hard on her. We normally do routine surgeries with nanobots, but, for this? This I had to do the old fashioned way.

The worms I removed ranged in size from about half a millimeter to ten centimeters. They have trilateral symmetry and, based on preliminary analysis of some chemical markers, belong to the dominant 'super-kingdom' of Eden."

There was a long pause.

"John?" George, leaning forward, took a moment to collect himself. "John, I...I apologize, but I have to ask. Are you *certain*? This is very difficult to accept. How could she possibly have been infected with an Eden life form? She hasn't even been to the planet's surface. And she hasn't been *anywhere* near any of our labs. How could this happen?!"

John raised a hand to calm George. "There's no need to apologize," he responded. "I have no idea how she came to be infected. But, yes, I am certain. We ran the analyses multiple times. I dissected one of the worms myself. It *does* have trilateral symmetry, and it *does* have an Eden organochemical suite. The worms seemed to have consumed Jian's fat reserves, explaining her leanness, and were starting to feed on her muscle tissue, explaining her lethargy."

"You said you had no idea how she became infected," Elke interjected. "but I don't like coincidences. She and Esteban both, apparently, experienced a reduction in body fat. You said these...worms were infesting Jian's fat, muscles, and organs. I can't help thinking of Esteban's...remains. To be safe, I'd like you to take similar scans of Natalia, Corinna, and Dani."

Dani opened her mouth to object, but quickly stopped. "Yes, please. I haven't noticed a drop in body fat, but I'd prefer not to take any chances."

Elke turned to Dani, "you named the individuals that spent a night, or at least an evening, with Esteban. I'd like the names of *everyone* he interacted with since we reached the planet. We should screen anyone that had frequent or extensive contact with him."

She turned back to John and asked, "okay, let's accept, as a working theory, that the worms are from Eden and that Esteban was also infected with them. Did they reduce him to what we found? If so, where are they now? Did they die when he died, or are they roaming the colony?"

John held up his hands and shrugged. "Those are the right questions, and we don't know yet. It seems logical to presume that the parasites survived him and consumed his remains after he died because we found droppings. By the way, this helps explain why the droppings were small." He paused, then added, "that may be good news. If they're small, we wouldn't expect them to attack anyone else. On the other hand, they may be able to infect others. We just can't know. I plan to infect some lab animals with the worms we removed from Jian so we can see how the infection progresses. Until then, I have very little information from which to speculate."

Elke sighed loudly, then clapped her hands together. "John, thank you for calling for this meeting. Please conduct the scans as quickly as possible so we can have the results for the next meeting. Given the rising threat level, let's reconvene in two hours, our originally scheduled time for the next reports. Does anyone have anything else to report? No? Okay then, we'll reconvene shortly."

▪ 19:02

Jenna sat brooding in a conference room. She had called an emergency meeting with select colonial business leaders and briefed them about Esteban. They were as annoyed with the news as they were frightened. She rubbed her temples, trying to stave off another headache.

"Is it possible that anything we obtained escaped?" she asked reluctantly.

Chang grumbled, "no, because we haven't collected anything. Understand that this is delicate. My people have been *gently* engaging with some of the landing party personnel, hoping to find someone we can convince to bring back something worth exporting. But let me emphasize 'gently'. If any of these people escalate these inquiries, the game's over before it begins."

Jenna sighed and leaned back in her chair. "Yeah, I understand, but it's frustrating. The game may not be over, but it's definitely suspended until we resolve this situation. All landing parties have been recalled and further expeditions suspended."

Chang shrugged, "this is a long term play. It's fifteen years to the next star system. A few days or weeks won't matter."

. . .

Two hours later at the designated regroup time, Elke entered her conference room and was confronted with exhausted faces.

David and Cecilia arrived moments after she did; as they took their seats, she asked David how he was holding up. He mumbled, "it's been an extremely long day." Elke nodded, but noted his fragility.

She looked at every face in the room, making eye contact with each of them, as she affirmed them all. "Yes, it has been a very long day for all of us. Tired people make mistakes. I'd like to hear what you've learned since our last meeting, and I want to know what tasks you plan to delegate to your teams so that you can get a few hours rest. We can reconvene again in the morning."

She turned to John first. "How is Jian?"

John leaned back in his chair and exhaled, rubbing his hand back and forth across the top of his head. His exhaustion and concern was very evident. "I'm no longer concerned about Jian, but she is still recovering from surgery. I expect she'll regain consciousness in the morning. We've continued to scan her body and have not found additional parasites, at least not yet. Now, you might be wondering how long we need to continue scanning her, and I wish I could tell you. I simply don't know. Once we identify the parasite, we might be able to extrapolate from the parasite's behavior in its typical hosts. Even then, we can't be completely sure. My concern is that the parasites we extracted varied widely in size, suggesting to me, the possibility that there may be more too small for us to detect. Bottom line?" He shrugged, "bottom line recommendation is to keep her in Sick Bay another week, scanning her body regularly, then let her leave but have her return for daily scans for a designated period. Best I can offer."

He leaned forward, then continued, "as to Esteban's other contacts. Dani's scan was clear. Corinna is thoroughly infected, and the parasites are larger than Jian's. Natalia is infected, but to a limited degree, interestingly. We only found small parasites, and only in her lower abdomen. As we speak, my surgical staff is removing their parasites. Afterwards, we will follow the same protocol I've proposed for Jian.

Now, the localization of Natalia's infection in her lower abdomen is suggestive, to me at least. We know the parasites infest the organs. We found small parasites in Jian and Corinna's urine and feces. Now, I'm speculating wildly here, but I suspect that some of Esteban's parasites infested his seminal vesicles and were passed to his partners when he ejaculated. If that is the case, and if Robert Esposito was a sexual partner, he may be infected as well. Unfortunately," he paused and made direct eye contact with Elke, "we cannot reach the *Lucky Strike* to warn them."

He sighed, then concluded. "Dani provided my team a list of other names, people that spent time with Esteban but were unlikely to be sexual partners, at least since we reached Eden. My team will be contacting these people to arrange scans. She also provided names of other potentially intimate partners for Jian, Corinna, and Natalia. We'll want to scan them, as well. Finally, I've attempted to infect some lab animals with the parasites; I'll let you know whether that takes. Esteban's case suggests we should see something within a week."

Elke nodded to George as John sat down. He wiped away a nervous sweat from his forehead. After a pause, he threw down his hands. "Y-you…you have to understand," he stammered with anxiety, "we have not been here very long. It's only been, been what, three weeks? It will take years, decades, maybe generations to fully catalog and understand Eden's portfolio of life forms. Thus far, all we've been able to do is collect samples and build collections. We've barely started that work and haven't even begun to try to map out evolutionary relationships, behavioral patterns, ecological structures, anything. We haven't even formally named anything yet. We…my team…I, we couldn't have predicted this, there was no way we could have known. There simply hasn't been time!"

"George!" Elke interrupted, but her tone was gentle. "George, we understand that. We know your team has been working very hard, and that this crisis requires your team to make profound shifts in your working theories and work approach. You're amongst friends. Please, just tell us what you've learned."

George nodded, then sighed and said, "thank you, Governor." After a moment, he flicked his wrist computer. One of the worms appeared.

"This," George pointed to the projection, "is one of the worms John removed from Jian. We sequenced its DNA and learned a few very interesting things. First, all of the worms from Jian's body were genetically identical. Like identical twins. I'm waiting to sequence the DNA of the parasites removed from Corinna and Natalia. If those worms are also genetically identical, it would reinforce the idea that they were infected by Esteban in some way.

As I mentioned, we've been collecting and cataloging samples. For every sample, we sequenced DNA, captured images, and noted observational data about the sample site. We cross referenced the DNA from the parasitic worms to see if we could find some matches." He touched his wrist computer and a round, red disk appeared to the left of the worm image. "This is a spore, approximately five microns in diameter. We've collected spores like this one all over Eden, deposited on the surface and high in the atmosphere. Its DNA is a match to the worms. Not identical, but the same species."

He flicked his wrist computer one more time, and a creature appeared to the right of the worm. It was similar to one of the creatures he'd displayed in the general briefing. This creature had a conical beak, a deployed sensory cluster, two tails, and six legs with two joints each. The first pair of legs were folded against the body and ended in a pair of powerful claws, and the other four legs were sturdy limbs to support the body.

"We collected this specimen near the tsunami site. It was approximately 10 centimeters in length. This is also a match to the worms. We have not observed a live specimen." He then brought up another image to the right of the displayed creature. It was the hide of one of the large beach scavengers from the tsunami site. Visible on the hide were a set of finger-sized, hard-shelled protrusions erupting through the flesh of the creature. Movement was visible within the protrusions. "We also found that the DNA matched the creatures we found within these hard-shelled protrusions. As of this moment, we're thinking these shells are something akin to a cocoon. This is a common pattern with parasitic wasps on Earth. The wasp larvae break through the skin of their hosts, pupate and metamorphose, changing into an adult wasp. We think these parasitic worms are doing something similar, producing these shells, then emerging as one of the creatures we found at the tsunami site."

He paused, then sat down, leaving all of the projected images up above the table. "Okay, those are the facts I have to offer. Everything I'm about to say is pure conjecture, but I think it makes sense. This," he pointed at the creature, then paused awkwardly. "Forgive me, I don't mean to make light of the situation. But my team found themselves referring to this thing as a 'kraken,' to simplify discussions. So, unless someone objects, that's what I'll be calling them as we go forward."

"Why 'kraken'?" Cecilia asked. "What's a 'kraken'?"

George smiled awkwardly. "It's a bit tongue in cheek, I'm afraid. A 'kraken' is a mythical sea monster that is, essentially, a monstrous octopus or squid. These creatures have three extensible tongues. Between the tongues and the claws, it reminded a member of our team of Earthly cephalopods, like squid and cuttlefish. They have tentacles that are reminiscent of these things' tongues; and two of the tentacles are designed to snap out and grab prey, like these things' claws. But they're small, so the team has been jokingly referring to them as 'kraken.'"

He paused, then continued. "In any case, the kraken belongs to the dominant 'super-kingdom' of life on Eden. We believe it has a complex life cycle with alternating generations, starting out as a haploid spore. In fact, we've been cross-referencing other Eden samples, and we're finding a number of the animals in this super-kingdom apparently have a complex life cycle with alternating generations including reproduction through spores and a worm-like generation. No idea how many are parasitic. The others we've found, so far, burrow in rotting material accumulating in forests, lakes, etc."

He paused, realizing he needed to explain. "I'm sorry for the complexity. I'll make it as simple as I can and, I promise, if you stick with me, you'll see why it matters.

We humans are diploid. All of our genes, our DNA, comes in pairs, one from each parent. Haploid means there is only one copy of each gene. There are organisms on Earth that are haploid. Usually, haploid organisms are one generation in an alternating generation lifecycle. We humans start as fertilized eggs, develop in the womb, are born, reach adulthood, and die. Somewhere along the line, we produce eggs and sperm that fuse as fertilized eggs. Hence, we are diploid, two copies of each gene. Our eggs and sperm are haploid, one copy of each gene. They are how we diploid creatures provide one copy of each gene to our diploid children. So, imagine if our haploid eggs and sperm did not simply fuse to form a diploid egg, but instead, grew into distinct haploid life forms with their own behaviors and ecological roles. And that, someday, they in turn produced gametes that would fuse to form fertilized eggs that developed into diploid us. We refer to this as 'alternating generations' and it's actually common on Earth. Plants have alternating generations, as do some parasites, jellyfish, and so forth.

Here's what matters. The spores are haploid, the worms are haploid, and the kraken are haploid. We *think* that when a spore lands on a suitable host, it infects the host with one or more worms. The worms are able to reproduce asexually, spreading through the host. At some point, the worms make their way to the host's skin and form something equivalent to a chrysalis or cocoon; after an indeterminate amount of time, they each emerge, not as butterflies, but as a haploid kraken.

Now, the kraken we collected had the same two specialized intromittent organs we've seen on other animals from this 'super-kingdom', suggesting that they mate and produce something that's diploid, which in turn, produces the spores."

He leaned back and shook his head, then began tapping a finger on the tabletop. "Again, this is all a lot of speculation, but it makes sense. It also raises the question of what the diploid generation looks like. It's a phenomenal evolutionary strategy for organisms on a planet with lots of ecological disturbances. The spores spread over the planet, parasitizing organisms that have endured a tsunami or wildfire. They capitalize on the disturbance, and, eventually, produce a hell of a lot of new spores as succession brings the environment back to maturity."

John asked, "what kinds of organisms endure wildfires?"

"On Earth, there are plants and trees whose seeds survive fires; some even require a fire to germinate. Some animals burrow underground. We expected to find the same strategy here. I expect we'll also find some equivalents to the animals we found at the tsunami site that have sufficient mass and ability to grasp the substrate to hunker down and let the fire wash over them. Remember, we found some trilaterally symmetrical animals whose dorsal segment skin is made of a very tough polymer. The molecular biologists are excited about the commercial potential of that polymer!"

He leaned forward, his speech rambling a bit as he thought it all through. "Where does that leave us? We think we're dealing with an ultimate pioneer: a life form that capitalizes on disturbances. The spore, on contacting a host, has some ability to 'design' the worms it produces to match the biochemistry of the host, and that's what's allowed the parasite to infect Esteban and the others. We believe the worms reproduce asexually, probably by splitting because they don't seem to have reproductive structures. We also believe they consume each other. We found semi-digested bits of worms in the guts of the worms we dissected. We think the worms prioritize consumption of body fats. Eden's fats may not be the same as Earth's fats, but, chemically, fats are fats, and a successful parasite keeps its host alive as long as it can. We think they then go after muscle tissue; again, amino acids are amino acids, which would have weakened Esteban to the point that he could not leave his bunk or call for help. When the host is no longer able to move, they metamorphose and emerge as 'kraken' that finish devouring their host. Then they disperse. Larger individuals prey on smaller ones; we found some of their tissues in their droppings. But, for the most part, they dispersed, presumably through the air ducts."

Elke asked, "why would they eat each other?"

"Speculation? It makes sense as a strategy. After a disturbance, ecological succession brings the environment back to its normal state. A pioneer's strategy is to, ultimately, produce as many offspring as quickly as it can before its life cycle is interrupted. The worms reproduced asexually as they consumed the host as rapidly as possible; then, as they exhaust the host, they consume each other. Not as efficient as one worm consuming the host, but overall the maximum number of the largest possible worms metamorphose in the least possible time. Then, as they finish consuming the host, again, they turn on each other. And remember, they're genetically identical. While individuals may not want to be consumed, evolutionarily, it's a winning strategy that results in the highest number of the most successful 'bets' on, eventually, generating the most possible spores. The individuals we see, of any species, are the descendants of the ones that placed the most successful reproductive bets."

He tilted his head to the side, processing another thought. "I'd guess… that they're in a race to reach sexual maturity before the ecosystem returns to its mature state. Maybe the kraken can't compete when the ecosystem matures. Maybe, in the mature ecosystem, there's a predator that eats young kraken before they can reproduce. And we can't know what 'sexual maturity' entails. But I would guess it's a function of size. We don't know how far they've spread, how many there are, how big they are now or how big they'll get. But I would guess they will all be aggressively seeking food to grow as fast as they can." He closed his eyes and sank back into his chair.

The conference room was silent.

Elke spoke first. "George, John, thank you. That's a lot of information to absorb."

She paused and blinked several times. She was exhausted and assumed her staff was too, but she couldn't show it. They needed her to be strong. It was a struggle, but she collected her thoughts.

"So, if I'm following, we have parasites that infected Esteban through some unknown way, possibly involving these spores, and he apparently infected his partners. The parasites left his body by turning into these kraken, consumed his remains, and are now roaming the colony. We should assume they're seeking food and, at some point, will reach breeding size, which will result in many, many more spores, which could infect other colonists. Do I have that right?"

George and John looked at each other, then nodded. "Do remember," George added, "that mating will result in a distinct, diploid generation that could be completely different. And I don't think the diploid form of the kraken will have a long lifespan. We've only been here a few weeks, but all of the individuals we've sampled from this group have been haploid. Our sampling has, admittedly, been limited, but we haven't encountered any diploid individuals. So, they could be short lived, or might be a parasite within the kraken's body, or any of a number of alternate possibilities. All we can say, with any degree of certainty, is that, once the kraken start mating, the risk to colonists increases exponentially."

She leaned forward and shook her head. The situation was starting to frustrate her, or maybe she was just tired. "Ok, well, instead of hunting some large predator that attacked and consumed Esteban, we're hunting some unknown number of smaller creatures that will infect other colonists if we don't find them first. I think I prefer the large predator threat, frankly. But we have to deal with the reality in front of us."

She looked up, "does anyone else have anything to report?"

Dani said, "we have not had any reports of any sightings. I can only presume that these 'kraken' are somewhere in the colony, but my team has seen no evidence of them. At least not yet."

Moira chimed in, "my engineers are researching what weapons we can fabricate, but we haven't started production because, well, we don't know what we're up against yet. We're considering handguns with rubber pellets that would pierce an animal but not the colonial drum itself. Looking at George's sample, I think that has merit. But I'll feel better if we could catch a couple of these things and better know what we're hunting."

She flicked her wrist computer and a schematic appeared of the air ducts in Ring 6, including the four broad conduits connecting them to the drum's core. She stood and began to point out specific sections: "after the discovery of Esteban's remains, you ordered the air ducts closed. I thought it would be useful to understand what that does and doesn't do. Each ring has four primary linkages with colonial life support, which is why we were able to isolate each ring's four quarters. Within each quarter ring, there are ducts branching from the colonial life support main to each room." She flicked her wrist computer again, and the adjacent rings appeared. She continued, "As you can see, there are ducts interconnecting the rings. We closed those ducts as well. That creates some minor problems with relative air pressures as people move between rings. Nothing we can't live with indefinitely.

But we cannot know how far the kraken spread before Esteban's remains were discovered. We don't know how much time passed, and we don't know how quickly they move. But here's what we *do* know. We know the air ducts are, unfortunately, the primary pathway for the spread of vermin through the drum. So, we, or I should say, Nikko's people, routinely place traps in the air ducts. They check those traps regularly to monitor the vermin population throughout the drum."

Nikko cleared his throat. "I had my team clear traps after the first meeting, and then right before this meeting. As we feared, we found that vermin populations are significantly reduced in some sections of the colonial drum, suggesting to me that the kraken have been feeding on the missing vermin."

Nikko nodded at Moira, and she tapped her wrist computer. A network of air ducts became highlighted in red. "As you can see," Nikko continued, "vermin counts were reduced through these ducts, indicating spread to adjacent scientific workspaces and living quarters in Rings 5, 7, and 8. There also seems to be a significant shift to the colonial life support core and to these ducts connecting to the agricultural rings."

The projection changed again, and the agricultural rings appeared. "Once we reached Eden and shifted these rings to the colonial drum," Nikko motioned to the projection, "we mixed the live soil we've been making on our way here with the sterile soil the drones placed in the broad agricultural rings they built for us. We encouraged the farmers to plant grass through those rings. We opened as many portals as we could between the rings to let livestock roam much bigger pens. As a result, we have broad swaths of grass with pens, coops, and so forth. We also have fields planted with a number of crops transferred from hydroponics to fields. Makes for happy goats, chickens, and pigs, and, I fear, happy kraken."

Elke's hands pressed together tightly. She rubbed her nose as she looked down at the conference room table surface. After a pause, she looked up. "We don't know how many kraken there are, but we have a pretty good idea where they've spread based on the vermin counts. We don't know how big they are or how big they'll get, but when they start reproducing, we risk seeing colonists and presumably livestock infected with more of these parasitic worms. Nikko, can you, based on the data between the vermin counts before and after Esteban, calculate the aggregate mass of the consumed vermin? Something to give us some idea how many of these kraken we're dealing with, and how big they're getting?"

Nikko nodded, "absolutely. I'll get the team working on that immediately!"

Elke looked at Dani and Jenna. "We've already informed the colonists that something attacked Esteban. Do we need to inform them that we're facing potential infections rather than some marauding predator?"

Dani said, "if you're confident that's what we're facing, that would be reassuring."

Elke shrugged and said, "at this point, I'm not confident at all. But, as you also heard, that seems to be the situation." She placed her hands on the table. "Okay, I think we need to inform them that, while they may be in danger, they're unlikely to be attacked. It's more of a potential infection. It's, what, 1am now?"

She sighed, then continued, "I don't think there's any benefit to waking them up. We'll revise our message, and they'll see it when they wake. Meanwhile, Dani, I suggest you get your team working on how to sweep for these things. Anything else?"

After a moment of silence and a few yawns, she decided there was nothing more to do for now. "Tired people make mistakes," she observed. "My gut tells me we're going to have some long nights before this is over, so I want everyone to get some sleep. Knowing you as I do, I'm sure you'll think of important tasks that need to be done. I want you to delegate them to your teams and get some sleep. If you need John to get you something to help you sleep, do it. Am I clear?"

She looked about the room, locking eyes with each of them, nodding and getting a nod in return.

"Get some sleep, grab breakfast, and be back here at 09:00. Good night."

▪DAY 23, 07:12
PLATFORM TIME

The creature continued walking through the green blades, gaining confidence as it grew and as it became increasingly familiar with its surroundings. Somehow it knew that it was approaching a size at which it would no longer be vulnerable, not even to its siblings. Soon, it would be large enough to breed, and could begin looking for a mate. Warm colors pulsed along its flanks.

The kraken could feel new vibrations and it could sense electromagnetic fields. It began moving towards them. It had feasted on more of the things covered in little gray threads. But it was growing rapidly and needed larger prey.

As it walked, it began to hear new, odd sounds. It strained its sensory cluster to see over the green blades. Not too far away, it could see prey. There were four of them. They were white, with two yellow legs and feet. The bodies were covered in something fluffy. They were walking in the green blades, around a long silver thing. Their heads moved forward and backward abruptly. Their heads had optical sensors, like the gray things it had been feeding on. Their heads had yellow beaks, but not like its beak. They stopped, occasionally, looked at the silver thing, and pecked.

It was excited. But the prey wasn't moving away from the silver thing; they seemed quite intent about it. Perhaps it held *their* food source.

The kraken matched color and texture with the green blades and waited. And waited. The prey did not come closer. After waiting, for an incredibly long time, it decided to move closer. It withdrew its sensory cluster, ready to retract it into its fold. It moved closer. And closer.

It was very close. It could see that the silver thing had something on it, and the prey were pecking at the things on the silver thing. They *were* feeding. That was very good. They would be focused on their feeding.

It moved just a little closer and waited. It began sending out electromagnetic pulses to confuse the prey.

Closer...come closer.

One of the prey moved close enough and the kraken shot out its claws. The left claw snapped shut about the fluffy white body and the right claw snapped shut about the neck just behind the optical sensors. The other prey squawked and ran away. The one it held was *strong*. It was hard to hold onto! The fluffy body was irregular; there were limbs along the sides that flapped, raising the prey upwards. The limbs strained as the prey tried to pull away, but the creature was too heavy and too strong for the prey. The panic and the squawking from the prey was so exciting!

It began to carefully pull the prey towards itself, as the limbs on the sides continued to beat down. It retraced its sensory cluster into its fold to protect it, relying on its lateral and electromagnetic sensors. As it drew the prey in, it could feel some of the internal structure giving way. It clamped down *hard* with its claws, and it could feel the prey's pain and panic as it did so.

Finally, as it pulled in the prey, it opened its beak and bit through the fluffy, white body. It was delicious! Such exquisite fatty deposits! New tastes in the meat! Red liquid flowed out, rich with iron, and waste products were voided. Its tongues lashed out and drew across the body, pulling off the fluffy white things that did not taste good at all, revealing the tasty flesh beneath. It bit down again and could feel the internal structure breaking; it made a wonderful sound. It stopped sending electromagnetic pulses so the prey's mind would clear. The panic and the pain and the fear from the prey was just *so* intoxicating. It wanted the prey's mind to clear so the prey would experience everything that was about to happen, and they could share in what was to come.

The kraken continued to slowly feed, fighting the urge to feed too quickly; it began exploring the internal structures of the prey, deciding what could be consumed and what needed to be left alone, for now, to keep the prey alive. It was interesting to discover the similarities and differences compared to the other prey.

As the prey weakened, it deployed its sensory cluster, and it aimed its eyes at the prey's eye, staring as it fed, slowly, deliberately, on the prey.

The prey stayed alive a wonderfully long time, but, eventually, its pump stopped pumping. As it felt the chicken's life slip away, it carefully and methodically dismembered the remains. It would not waste any of it.

As it finished, the kraken scanned the area, looking for the other prey. It could feel them nearby. It decided to wait, right here, by the silver thing with the food. It blended with the background, started sending more electromagnetic pulses, and waited for them to return.

▪ 08:55

Elke carried a coffee mug with her as she walked through the corridors. She'd slept, but not well. This wasn't supposed to be happening. There wasn't supposed to be a crisis, or casualties, no existential risk to her colony.

She shook her head as she walked. *No time for this. Too much work to be done.* People needed leadership; they needed support and encouragement. *We can recover from this, stay strong.*

She'd reached her conference room. She paused, collected herself, and entered. Everyone else was already in their seats. She could see they were stressed as well. She took her seat, scanned the room, and took one more sip of her coffee.

"Good morning. I hope you each were able to get at least some rest. We have too much to do, not enough time to get it done, and an awful lot of people counting on us to get everything right the first time." She paused to sip her coffee again. "But I have complete confidence in this team. You have the skills, as well as the experience. You have the insights and the wisdom. Let's get to work."

She turned to John and asked, "how are your patients?"

John nodded tiredly. A normally very well-kept man, he was a bit disheveled, a true sign of his tiredness. "Good morning to you too, Elke." He sighed, "Jian is awake. Weak, but awake. She's shaken, of course, after learning about her infection and Esteban's fate. But she's doing ok. Scans remain clear. We hope to have her on her feet soon. Corinna and Natalia are both recovering from their surgeries. Natalia's was not as invasive, so we expect her to have a smooth recovery. Corinna's surgery was the most extensive, so she won't be waking up for a while. The parasites had thoroughly invested her organs, consumed her body fat, and were attacking her musculature. I suspect she would have shared Esteban's fate sooner rather than later. She'll need ongoing treatment to repair the damage."

He paused, then added, "let's talk about the other contacts Daniella identified. We've scanned everyone that had contact with Esteban. As we've noted, three of his intimate partners were infected, one was not, and one cannot be tested because he's aboard the *Lucky Strike*. Now, we've scanned all of the other intimate partners of the three infected partners, and they are all clear. We've also scanned all of Esteban's other, non-intimate contacts, and they are also clear.

Where does that leave us? We have one individual infected after being on the planet's surface, and three individuals infected after intimate contact with that first individual. No no one else has been infected. All of the worms removed from the three infected individuals are genetically identical to the alien genetic material in the droppings found in Esteban's quarters. That, plus the infection pattern noted in one individual, Natalia, suggests that the parasites infested Esteban's reproductive system, probably his seminal vesicles, and were passed with his ejaculate. The absence of any infections resulting from intimate contact with infected females suggests there may not be a comparable way for females to infect their partners. We still don't understand how Esteban was infected, but we do have a reasonable model for how he passed on the infection."

After a pause, Nikko cleared his throat, then flicked his wrist computer. An environmental suit appeared over the table. He said, "last night's conversation was pinging in my head, so I had my team find Esteban's environmental suit and pressure test it."

He touched his wrist computer, and the image expanded, zooming in on the ankle of the right boot. He stood and pointed at a small black spot. "Sure enough, there's a tiny hole on the ankle of the suit. We think one of these spores landed on the suit, managed to pierce the suit, and infect him."

Cecilia gasped. They looked at her and she dropped her head, embarrassed. "I'm sorry. I just remembered, I noticed that Esteban was scratching his ankle a *lot*. I remember noticing it, before one of the briefings. I should have encouraged him to go to sick bay, but I wasn't sure it was my place to say anything. And it was just an itch, I mean, how could I have…I mean, what if I'd said something? Might he still be alive?"

John leaned forward and said, "there's no way you could have known. And there's no way to determine how far along the infection was. It might not have made a difference, so please don't blame yourself."

She nodded, unconvinced.

Nikko continued, "I can't know whether the itching was from the spore piercing his skin, or from some of the disinfectant getting into his suit through the hole. We use some nasty stuff to disinfect when they come back from the planet's surface." He returned to his seat. "Ideally, we'd get more of the spores and do some testing, but I don't see how if we're not going to the surface. I'd also like to know more, George, about the critter you mentioned with a tough polymer hide. I want to know how similar that polymer is to what we use for the environmental suits."

Elke tapped her coffee mug nervously. "So, if we resume landing parties, we might see more infections?"

Nikko nodded, "right, and we need to consider the possibility that other people that have been to the planet may also be infected by a kraken or something else with a similar life cycle. My team is pressure testing every environmental suit that's been to Eden. We disinfect, of course, when a landing party returns, but we only pressure test before they head to the planet."

Moira jumped in. "It's reassuring to have a credible theory for how he might have been infected, but the possibility that the polymer suits could be compromised is a significant concern. It's a very tough polymer. We use that polymer for a number of things. If these spores can compromise that polymer, we cannot have those spores get loose in the colonial drum."

"Exactly," Nikko pointed towards her in agreement. "We have to find and kill these things before they generate more spores. Not only might they infect other colonists, they might compromise the integrity of the colonial drum."

There was a long silence.

Elke stiffened her back; the coffee was not working as quickly as she had hoped. "Nikko, continue to pressure test the other suits. If you find another hole, inform me immediately. Dani, I'm thinking that we should presume the worst. If Nikko finds another compromised suit, let's immediately locate the owner. If they're not in their quarters, have them report directly to John in Sick Bay. If they're in their quarters, lock them down. Close their air ducts; lock their doors. I don't want any more of these things escaping through ductwork."

Dani nodded, "done."

Elke continued, "Moira, Nikko, I need ideas for dealing with the spores. I want the air monitored to detect them. I want options for filtering them out of the air, and for isolating ring segments where spores are detected."

They hesitated, then Moira turned to Nikko, "we can come up with a detector and we can start monitoring the air throughout the colony. We can come up with a filter, too, I'm sure."

Nikko smiled nervously in response. "I'll get you a list of the duct intersections. That will give us a tally for the filters we need and their sizes. We'll also verify the emergency bulkheads are working."

Moira said, "we should consider restoring the lateral ring walls in the agricultural rings to reduce the scope of any structural failures."

Nikko winced. "That's a *lot* of work and will set back some of our agriculture. How about we compromise. I'll have some of my people work with the farmers to see what we can do about putting *some* walls back. We can do a risk assessment regarding isolations in the event of a breach. Deal?"

Elke nodded, then turned to Dani. "Anything to report? Any sightings?"

"No sightings," Dani replied. "We're patrolling the living spaces as best we can, but we aren't staffed for this, let alone sweeping everywhere else. We're supposed to be helping inebriated colonists get home without hurting themselves, not hunting kraken."

"Understood, Dani, and thank you. Nikko, any chance any of your operations team have applicable experience? Jenna, do any of the colonists have a security background?"

Nikko said, "not that I'm aware of, but I'll check personnel records. If nothing else, we can provide support."

"I'll check as well," Jenna said.

Dani nodded to them both, "thank you, I appreciate the support."

Elke turned back to Nikko, "any further signs of our unwanted guests?"

Nikko said, "vermin traps are coming up empty in a number of places. The insect counts in the air ducts are steady, depleted, but steady. But now the rodent counts are down, including in the working and living areas. This suggests that the kraken have grown too big to be interested in insects. We have some complaints about missing small pets and livestock: cats, small dogs, and some chickens. But we were getting such reports before all this happened, so it's hard to say whether it means anything.

We also have vermin traps in the plenum areas, and we started checking them as well. Those trap counts are also down, suggesting our guests are roaming that space. That's a concern, and a lot more space to sweep."

Jenna asked, "what do you mean?"

Nikko looked at Moira, who said, "go ahead, you're on a roll."

Nikko smiled and flicked his wrist computer. A model of a colonial ring appeared.

"Each ring," Nikko explained, "is 500 meters in diameter and twenty meters wide. Well, that's the ring we transferred from the *Lucky Strike*. The extra agricultural rings built by the deep space probes in preparation for our arrival are the same diameter, but they're 500 meters wide. They were built in place, so they didn't need to be designed to be moved between colonial and deep space vessel frames, and that broad width really helps our farming until we build a dome. It's a trade-off. That's also a big space to lose in the event of a breach, but we've never lost one in any of our colonies. The only breaches we've had have been small leaks; worst case, we lose a little air until the breaches are fixed."

He pointed at another circle inside the colonial ring model, then continued. "These agricultural rings are designed for crops and livestock. In addition to the broad space, they have high 'ceilings', this inner ring. It's thirty meters above the outermost rim, with structural supports in addition to being attached at the sides. That lets us manage lighting to simulate dawn through dusk as well as seasonal day length variations."

Cecilia raised a finger and asked, "where did the soil come from?"

Nikko turned and smiled. "We touched on this last night. It would not have been reasonable to bring that soil with us, and we can't use Eden soil without sterilizing it. The deep space probes that built those rings provided the soil by pulverizing rock taken from Eden's moons. When we arrived, we infused that soil with live soil we made during the crossing, as well as a microbial mix. When the mixture was healthy with microbes and suitable nutrients, we seeded it with grass. Make sense?"

When she nodded, he returned his attention to the model, which changed to highlight a series of inner rings.

He continued, "here in the command, residential, research, industrial, and commercial rings, we have an inner ring only five meters above us." He pointed up, "above that is another twenty meters of what we call the 'plenum' space. As the colony grows, while we're building a dome or another orbiting frame, we have the option of adding living and working space one, two, three, or four more levels above us."

Right now, though, that means we have only twenty meters of open space above this ceiling. That space has very little in it, just the structures that would, eventually, support the flooring, ceilings, and other infrastructures necessary to convert it to living and working space. It's open, it's dark, it's very rarely accessed. It is pressurized, served from the same air duct mains that support the space we are using."

He flicked his wrist computer and the ring model disappeared. "Above that we have unpressurized space with structural supports. We also have engineering facilities, including the life support core, which is, of course, pressurized. There are some other engineering and lab spaces that don't require pressurization. I don't think we have to worry about our guests getting into the unpressurized areas."

As he sat down, Cecilia let out a groan. "Shit."

Dani, nodding, said, "eloquently put. Again, we've never had to deal with anything like this before, so we're really not designed for it."

Elke nodded, "it is what it is, so we'd better figure out how to sweep those spaces."

"We're working on it," Dani meshed her fingers together. "My understanding is that there are some lights we can turn on. I'm more concerned about our ability to physically navigate the space; there are footholds designed to support the operations team should they need to convert it to living or working space, but they're sparse." She glanced over at Nikko, "definitely an area where we can use some support from your team."

"Absolutely. Let's work on it when we're finished here."

Nikko took a deep breath, and his face became very serious. "As I said, we've continued to check vermin traps and are now tracking the traps in the plenum. That gives us a rough estimate of the amount of food the kraken have consumed and where they've been. We cross-referenced that with livestock growth charts for some rough, very rough, estimates. By our calculations, we believe there are now about eighteen kraken, and they've reached about 2kg by now."

After a pause, George let out a low whistle, "that is an *impressive* growth rate."

Nikko's eyes widened, "I know. I had them check their math a few times. And if the missing pets and livestock are real losses to the kraken, they may be bigger than that."

George frowned, "even your more aggressive growth projections leave them a little small for taking pets and livestock."

Elke asked, "do we have any basis to estimate how big they need to get to breed?"

"No," George shrugged. "The only way we could get that kind of data would be to resume landing parties and observe what's happening at the tsunami succession site. And I completely understand and support why that's *not* an option."

He paused, then asked, "how about drones? They're not designed for that kind of detail, but we might be able to play with them and see whether they can be adapted."

Moira nodded thoughtfully. "I have some folks that could work on that with you."

Elke tapped her mug, encouraged and more awake. "Please get on that, both of you." She turned to Moira. "Weapons?"

"We've made some of the rubber bullet handguns," Moira said. "They won't pierce the drum. Nikko, can you have some of your people oversee their distribution to Dani's security team? As soon as you can verify whether they're effective, we'll make more or come up with something else."

Dani asked, "could we also make some rifles? Or scatterguns? Handgun precision may not be adequate in the plenum."

Moira nodded, "we'll get to work on that."

"Dani," Elke took another sip of coffee, "I need a search and destroy plan. Do your best to capture some of the kraken, but we need to eliminate these things as quickly as possible."

George asked, "any chance of using drones to patrol the plenum? They could be equipped with more sensitive sensors than our eyes."

"They're really not designed for such tight quarters," Moira said, "but we could look into what that would take." She looked at Elke, "we're going to need to prioritize, Governor. I have engineers working on production and prototyping of weapons, spore filters, spore detectors, and now drone modifications. We can fabricate nearly anything, but we have a limited number of fabricators. If you expect demand to continue ramping, I'll need to commit one of the fabricators to making the parts for additional fabricators. But that would impact our ability to produce filters and weapons."

Elke meshed her fingers together, contemplating their next move. "Right now, the spore filters are our top priority, then weapons. The drone modifications are important too, but not as time critical." She looked around the room, "anything else?"

Silence.

She stood up and grabbed her now empty coffee mug. "Let's convene again in four hours. Dismissed."

- 12:41

Elke sat at her desk. She'd just completed and transmitted her latest status report.

Obviously, they were on their own. They were nearly fifteen light-years from the nearest colony; any help dispatched today would take fifteen years to arrive. She could keep deep space command informed, but there wasn't much they could contribute. Nothing like this had ever happened before; nobody there had any better ideas than the people on site.

And there were the inbound ships, launched years ago. They could expect one ship per year, each carrying 1,000 colonists, for the foreseeable future. No choice, they were already in flight and could not be reached at their current velocities. Even if they could be contacted, they could not turn back or be redirected. They were coming.

They needed to get this infestation under control, fast. They needed to prepare for growth, fast. Like any new colony, they had more than enough room to double, even triple their population; but, given the inbound cadence, they needed to be planning for further expansion soon. If they couldn't get the infestation under control, if spores compromised the integrity of the drum, an awful lot of inbound colonists were going to be in a great deal of trouble.

At least they didn't have to debate whether humans could walk the surface of Eden without environmental suits. That was out of the question. Instead, they'd be developing new polymers for a new generation of environmental suits the kraken spores couldn't penetrate.

She stood and walked towards her kitchenette to brew another cup of coffee. Suddenly, she received a message from Nikko.

"Governor, we've found seven compromised environmental suits. Six people are on their way to Sick Bay. The last one is not answering. He was last on the planet's surface a week after Esteban. We've locked down his quarters per your instructions. Dani and George are meeting me there. Ring 5, Section E, Room 34."

"On my way."

■ ■ ■

As she approached, she saw Nikko, Dani, George, John, and four people she did not recognize. She guessed two were from the security detail. Both had handguns and one had what she presumed was a prototype rifle. The other had a standard security pack. She guessed the other two were with George; they had a suspensor pallet of specimen containers. John had a shoulder bag medical kit.

Elke said, "I'm glad you're prepared, but I'm hoping we're wrong."

Nikko nodded, then asked, "how do we do this?"

"I'm not worried about Tam," John commented. "He's either fine, or dead at this point. I'm worried about little kraken."

George pointed to his supplies, "we brought barriers; I suggest we set them up to block the corridor on either side of the door. We have heavy duty tongs for grabbing kraken."

Elke looked at Dani, who nodded. "Very good; set up the barriers."

George nodded to his team and they set to work. The barriers were a meter high and snapped in place across the corridor. When they were done, they picked up their tongs. They nodded, and George turned to Dani, "we're ready."

Dani turned to Elke, who gave a grim nod in command.

"Opening."

Dani touched the access controls, and the door became transparent. They stepped through to enter.

It was a standard room for a junior science specialist, with a small ante-room consisting of a table, chair, and communications console. A bunk room was partially obscured by a lavatory and shower.

Dani shouted, "lights, full!" As the lights came on, they heard scurrying sounds.

George pointed, "there!"

One of his techs stepped forward and grabbed a small kraken with a pair of tongs. It was almost ten centimeters in length, about the size of a hand.

As the tongs closed on it, it snapped its claws to grasp the tongs and began flashing deep, dark red colors along its flanks. George held out a specimen container and the tech tried to put the kraken into it, but the creature kept its claws clamped on the tongs and bit at them with its beak.

George finally said, "put the tongs in with it! We have more tongs." The tech dropped the creature into the jar and yanked his hands away as George shut the lid quickly. The kraken went berserk within the jar; George set it down and backed away.

The other tech stared at a corner of the room, then pointed. "I swear something moved over there."

George approached, reaching with the tongs; as he touched the spot, claws flashed and dark reds began flashing. He closed the tongs and stood, holding the kraken, and put it and the tongs into another specimen container.

John exclaimed, "well now we know why we haven't found any of them on our sweeps! George, had you observed such camouflage before?"

George stared at the specimen container that had been returned to the suspensor pallet. The kraken, having released the tongs, matched the color and texture of the container's floor. "No," he answered over his shoulder, "but you can see how effectively they match their background. Remember, we haven't had that many hours on the surface; we may have missed them." He pointed at the kraken, "it's not only matching the colors; it's matching the textures. And so quickly! Amazing."

"Everyone…" Dani called from the bunk room. Her tone spoke volumes. "You need to see this."

They walked into the bunk room. George grimaced. The others stared, struggling to comprehend what they were seeing.

Tam was on his bunk. He was lying on his side, wearing briefs. His face was frozen in an agonized stare. His skin was loosely draped over his bones; very little of his musculature remained. In some places, his skin was shredded. In other places, his skin rippled from the kraken moving under his skin. Very little blood stained his sheets. Small droppings, like the ones in Esteban's room, were scattered on the bunk.

A number of hand-sized kraken feasted on his remains; bright, red colors flashed rapidly across their bodies as they ignored the intruders. One was holding a fingertip in its claws, its digestive organ everted against the fingertip. About a centimeter of the fingertip had already been consumed. Two of the other fingers on that hand were completely free of flesh. Another was perched on his hip, claws holding folds of his papery flesh, tongues rasping muscular tissue through a tear in the skin.

Three kraken, with their feet on the sheets, inserted their heads and claws through tears in the skin of his chest and belly, presumably feeding on his organs. Another kraken, standing on the side of his head, was holding his ear in its claws and was rasping tissue from the ear.

One of the kraken was on the sheets, grasping his lower lip with one claw and his tongue with another. It rasped its tongues against his lip in a strange, grotesque kiss.

Dani sighed and said, "I don't know where to begin."

Tam groaned.

George muttered, "oh, dear God, he's *alive!*"

John leapt forward, putting a hand to his neck to check his pulse. Elke cried out , "George, get your team in here and remove every last one of those damn things. NOW!"

As George's techs stepped forward, John rushed over to Elke and Dani. "His pulse is weak and irregular," he spoke in a hushed and urgent tone. "I can't believe he's alive, but I'm not sure there's much we can do for him. I'm not sure he'd survive being moved, but he'll certainly die here."

The techs, with their tongs, began pulling kraken from Tam's body; but the creatures would not let go, and the tissues they were grasping with their claws came away with them. Tam groaned again.

Elke whispered, "can you at least anesthetize him?" She looked over John's shoulder at Tam. For the very first time, a shiver ran down her spine.

John rummaged through his medical kit. "I can try." He pulled out an ampule and pressed it against Tam's neck.

"We have a stretcher," Dani commented.

"If one of you will help me, we might be able to lift him with his sheets and place him on the stretcher."

Elke grabbed the sheets by Tam's feet as one of Dani's techs unfolded a stretcher from his pack. The tech with the rifle put it down. John grabbed the sheets by Tam's head.

He called out, "on three! One, two, three!"

They lifted Tam, shocked by how light he was. John guessed his weight was below 30 kilos. As they placed him on the stretcher, a kraken tore its way out of his abdomen. George grabbed it and threw it in a container.

Elke pointed at one of the security techs. "You help John carry him to Sick Bay! If you see another one come out of him, you knock it to the ground and stamp on it!" She pointed at one of George's techs and said, "you follow them, and you scoop up anything they leave behind! Now, MOVE!"

As they were exiting, she turned to George and his remaining tech and spoke through gritted teeth: "I want *every, last, one* of these little bastards caught."

▪ 14:00

As Elke entered her conference room, the silence was deafening, but understandable, of course. They needed time to process. But they also need to reclaim their colony.

She sat down and asked, "John, how is Tam?"

He kept his eyes down at the table. "Tam didn't make it to Sick Bay."

Silence.

Elke spoke regretfully, "I'm sorry to hear that. How about your other patients with compromised environmental suits?"

"Five are clear," John said. "Nikko was kind enough to tell us where the environmental suit holes were. We were able to confirm skin irritations at those locations but no infestations; presumably, a spore was able to compromise their environmental suit but was unable to infect them with parasites. The sixth was not so lucky. Her name is Emily and she is doing well. Weak, but doing well, considering. I don't expect her to be back on her feet for quite some time. We operated and were able to remove her parasites. She had very little remaining body fat and they were attacking her muscle tissue. They were larger than Jian's or Corinna's. I'm guessing here, but I'm thinking they would have, what? Matured? Encysted? Encysted, I think that's a better term. Anyway, I think they would have encysted soon. We're lucky we got to her when we did."

George said, "I agree with your assessment. We dissected the parasites you removed. There's a lot we don't yet know, but some of the internal structures were different than we'd seen in the other hosts. I'm speculating, but I think these structures have to do with preparing for metamorphosis; the process by which they transition from worms to kraken. Also, we checked their DNA. The parasites you removed from Emily are unrelated to the kraken we collected from Tam's room, and neither are related to the parasites removed from Jian, Corinna, or Natalia."

John added, "we were able to confirm, through post-mortem examination, that Tam's reproductive system was infested with the parasites. Infected males can pass the parasites through their semen."

"We're researching Emily's and Tam's contacts," Dani stated. "I'll have a list for screening by our next meeting."

Elke nodded and asked, "George, why weren't the others infected?"

He shrugged, "remember our speculation that we're dealing with an ultimate succession pioneer. Spores are cheap and they're raining down all the time. Most of the time, they fail to make contact with a potential host. When they do, and when they're able to get through the potential host's hide, they need to match the organochemical suite of the host.

On Earth, this would not be necessary; everything shares the same biochemistry. Here? Here, there are five distinct sets. Someday, we'll have time to analyze these spores and determine whether they are produced with random biochemical suites and just happen, by pure luck, to land on compatible hosts. Or perhaps they're designed to sample the potential host's biochemistry and adjust parasite biochemistry to match what they encounter. If I were to bet based on what we've seen? We have, what, eight compromised environmental suites and three infected patients? I'd bet that they sample and design."

He sighed, then concluded, "it'll take time to research that, but that's what I bet we'll find. But it doesn't really matter right now."

Elke nodded, "agreed. What have you learned about the kraken themselves?"

He flicked his wrist computer and one of the living kraken appeared over the table in a specimen container. It was hard to see. It matched the colors and textures of the container perfectly and was not moving. "Here's our guest. We collected nineteen of them from Tam's quarters. It took a while; they're hard to spot when they're camouflaged. We're processing the droppings we collected to estimate how many may have been cannibalized. We also can't tell whether any escaped through the air ducts before we shut them, but we're thinking they wouldn't start consuming each other or leave while the, uh, host was still available."

He became extremely uncomfortable. "In the absence of other data, we can only assume something on the order of nineteen or twenty have spread through the colonial drum from Esteban. That matches Nikko's projections based on the vermin traps."

He paused, then flicked his wrist computer again and a different kraken appeared. Colors flashed slowly along its flanks as it explored one end of a larger container. As they watched, a cricket was introduced to the opposite end of the container. After a moment, the kraken stopped moving, then effectively disappeared, matching the colors and texture of the container's substrate perfectly.

George paused the image. "Note that, when the cricket was introduced, the kraken's sensory cluster was aimed at the corner of the container in front of it, yet it quickly noticed the presence of the cricket. We think it detected the cricket with another sensory modality, and we've initiated a series of experiments to explore this. We *know* that many of the organisms in this 'super-kingdom' have a series of sensors, along the boundary of the dorsal and side segments, that consist of a nerve cluster in a mucous filled pit. We think these sensors detect vibrations in the air, like extremely sensitive and directionally precise motion sensors. There are also some internal organs whose purpose we have not yet ascertained, but one of them is highly enervated. So many nerve connections between this organ and the brain may indicate it's a sensory organ, but we're not sure."

Motion resumed. As the cricket explored its enclosure, the kraken remained motionless. Its sensory cluster was exposed, but its supporting limb was folded to hold the cluster close to its protective sheath. After a few minutes, as the cricket continued exploring one side of the container, the kraken very slowly moved sideways to position itself in the cricket's path.

The cricket paused periodically, using its antennae to explore, slowly approaching the kraken. When it was within a body length of the kraken, the claws flashed out blindingly fast. One claw grasped the cricket just behind the head and the other clutched the middle of the abdomen. It held the cricket firmly despite its efforts to break free and slowly extended its sensory cluster to inspect the cricket. It abandoned its camouflage, and the colors flashing along its flanks resumed. They were bright and rapid. After a few moments, it slowly pulled the cricket closer to itself. It opened its beak and bit through the abdominal wall, then dragged its tongues over the insect's surface. Holding the insect with its tongues and claws, staring into its eyes, it everted its digestive organ into the wound it had made in the cricket's side. As it did so, the cricket's struggling slowly intensified.

George froze the image again. "Let's look at that again," he said. He flicked his wrist computer, and the image was refreshed back to the movement of the cricket while the kraken was camouflaged, just before the attack. The kraken remained motionless, but the cricket moved extremely slowly. "We are now watching at 1/1000 speed."

As the cricket took one more step forward, the kraken suddenly lifted itself above the substrate to the full height of its four walking legs. The claws were attached to the body behind the beak, with the first segment of each claw folded inward under the body, the second segment folded backward, and the final segment, ending in the claws themselves, pointed forward. As they watched, the claws flashed forward, the first segments pivoting outward to widen the grasp; the claws opened and then snapped shut about the prey. The body settled down with the four walking legs braced forward to anchor the kraken against the struggling of the prey.

"That's how it hunts," George said, "largely a sit and wait predator, relying on its camouflage until the prey gets within range. The fastest striking predator on Earth is a marine creature called a mantis shrimp, achieving twenty-three meters per second: fast enough to boil the water immediately surrounding its claws. The strike you just watched was twenty-five meters per second and transferred enough energy to the prey's body to mortally wound it. From what we've observed, strike speeds vary; the fastest we've seen was twenty-seven meters per second and the slowest at fifteen meters per second. Still blindingly fast. The claws are powerful; they're capable of considerable pressure. The claws are designed so the kraken can lock them against the struggles of their prey. The claws are also equipped with an array of sensors; it can 'taste' the prey through the claws, sense textures, things like that. It can also deliver a powerful bite with that beak; we're quantifying that in the lab."

David asked, "have we seen anything like their camouflage before?"

"Yes, we have!" George responded. "Our tsunami team reported observing a large predator taking an herbivore near one of the blinds. While stalking, it matched the colors and textures of its surroundings. We've seen this on Earth, too. Chameleons are famous for their ability to match background colors. Cephalopods, uh, octopus, squid, and cuttlefish, match background colors and textures. Their skin is equipped with chromatophores and papillae that allow them to rapidly and precisely manage their skin color and texture, coupled with sophisticated eyes, brains, and nervous systems. They're amazing to watch; I recommend searching for videos of them. Now, we haven't investigated how kraken do this, nor have we attempted to quantify how their abilities compare to Earthly cephalopods. But, I gotta tell you, what we're observing here is damn impressive!"

"And why didn't the cricket see it? I mean, sure, the kraken matched the color and texture of the container, but it was still relatively big on an otherwise flat surface."

"Excellent observation…and we don't know, at least not yet. Crickets don't see particularly well. It will be interesting to see what happens when it's big enough to take a mouse. Which shouldn't take long, by the way. Their growth rates are impressive."

George pointed at the kraken, which was continuing to feed as the cricket struggled. Its eyes were still aimed at the cricket's eyes, and it had repositioned its claws; periodically, it used its beak to extend the wound in the cricket's body. "Something else to note about our friend here. It's a preliminary observation, but they seem to like to keep their prey alive while they're feeding. It sure seems to be making an effort to consume the cricket slowly and selectively, to avoid killing it as long as possible."

Jenna raised an eyebrow in alarm. "I thought predators killed their prey as an act of mercy?"

"Predators don't care about their prey," George responded, "at least not usually. Some predators dispatch their prey to subdue it, not as an act of kindness. A lion doesn't want an antelope breaking free or even injuring it while it feeds. These guys have their prey effectively under control, and they don't appear to feel a practical need to dispatch them. What's disturbing is that they seem to make an effort to keep the prey alive as long as they can. Pure speculation? It's an artifact of keeping their hosts alive as long as possible during the parasitic phase. As I said, that's pure speculation. Once the prey dies, though, they consume whatever is left quite efficiently."

He flicked his wrist computer again, and the image vanished.

More silence.

"Thank you, George," Elke said. "We need to learn everything we can about these creatures."

She turned to Dani and asked, "do you have a search and destroy plan ready?"

Dani answered, "right now, we're configured for rapid response. Between Nikko's operations team and Jenna's colonists, we've identified enough people with weapons experience to boost security from ten to twenty-seven. I've organized them into seven teams of three and three teams of two, with one of my people in charge of each team. The teams are working in two overlapping shifts, so at least five teams will be on call at any time, and they are positioned as follows."

She flicked her wrist computer and a schematic of the drum appeared, with the science, residential, and agricultural rings highlighted. She continued, "one team to the science rings, two to the residential rings, and two to the agricultural rings. That's still a lot of territory to cover, but it's the best we can do. Moira provided handguns for each individual and a rifle to each team. I'd like enough rifles for everyone. I'd also like to explore other possible weapons that don't require as much precision: flamethrowers, perhaps, or a scattergun with rubber pellets that won't pierce the rings."

Nikko frowned. "I can see the appeal of flamethrowers, but I'd be concerned about their use on an orbiting platform."

Moira leaned forward to intervene. "We'd need to equip them to coordinate with our fire suppression systems. They'd need to deactivate, or perhaps delay, fire suppression long enough to be effective but not so long as to put the platform at risk. We'll play with it."

Elke nodded, then said to Jenna, "thank you for contributing to the security team."

Jenna pursed her lips, as if she tasted something sour. "These people were volunteers. There are other colonists that are demanding more *action*: parents concerned about their children, farmers concerned about their livestock, people chafing at restrictions on their movements. I'm sympathetic to their perspective."

Elke looked at her for a moment, then clasped her hands together. "Speak plainly; we're family here."

The mood in the room was tense; no one moved. It was as if they were the only two present.

Jenna leaned forward, and spoke somewhat brittlely. "There are colonists that are unhappy with the situation. They don't…trust your teams to keep them safe. They want to take initiative, to protect themselves before someone gets hurt. I don't blame them."

Elke also leaned forward, then said, rather gently, "Jenna, I can understand how they feel, and I would be happy to meet with them anytime, anywhere. I'll listen to their concerns and I'll answer their questions." She paused, then added firmly, "you've been in every meeting. If you think we're not being aggressive enough, then speak up. If you have better ideas, speak up."

Jenna looked down.

Elke continued, "if you have colonists that are willing, and capable, to help, Dani will put them to good use. But let us be perfectly clear: right now, I need every one of Dani's people, including the colonial volunteers, focused on eradicating these kraken. We cannot spare even one to chase vigilantes around the colonial drum. And any colonist that pursues these things on their own is endangering himself and the other colonists."

Jenna looked up and Elke locked eyes with her.

"Is that understood?"

Jenna paused, then nodded.

Elke nodded back. "Excellent. Once again, I understand their fears and I'm willing to meet with them, anytime, anywhere."

She shifted her gaze to Moira, "how are the spore filters coming along?"

"Challenging," Moira admitted. "As George explained yesterday, the spores are only five microns in diameter. Filters capable of capturing them don't let much air through, at least not as much air as we normally move through these ducts as part of our life support system. We're experimenting with different designs, including something that would mount diagonally to increase the total surface area that air would move through. We have initiated monitoring the air for spores; however, nothing has been detected yet.

I expect these next few days are going to be challenging, but I have every confidence that we will clear the colony. Once we do that, we're going to need to resume our landing parties. To that end, I've tasked a few of my people to come up with an alternate polymer for the environmental suits. Other polymers have been explored, and we can fabricate suits with these polymers once we're confident the polymers are impervious to the spores."

Cecilia nodded enthusiastically. "Thank you, yes, we'll need to resume surface explorations as soon as possible. Could we add a pressure test on return? Then we'll know whether landing party personnel have been exposed."

"That's a logical suggestion," Elke pondered, "but, please, not yet. The safety of the colonists and the integrity of the drum are our paramount concerns. Remember, there are a number of inbound ships with additional colonists already in flight. We must be ready for the colony to grow by about 1,000 colonists per year for the foreseeable future. We're already behind our accelerated expansion schedule, and we're falling further behind every day. We need these kraken captured or eliminated ASAP.

Moira and Nikko, please get the weapons distributed and the spore filters deployed. I'd like to see prototypes for the additional weapons. Dani, once your teams are outfitted, we may have all ten teams conduct a search and destroy sweep to clear the colony rather than staggering their shifts. And we need to take the plenums into account. Think about it and get back to me. And get John those contacts to screen. Anything else?"

After an empty pause, she stood.

"No more deaths. See you in four hours."

▪ 18:07

The kraken walked slowly and cautiously forward. It had roamed the fields of strange green blades and feasted on the things covered with gray threads and then the things covered with white fluff. They were exciting and delicious but now they were too small.

It needed larger prey. It was sensing electromagnetic fields and vibration patterns that suggested something larger, but the somethings were in a strange formation with straight lines and right angles. It was unsure about whether it was safe to approach. True, it was getting larger and larger, and soon it would no longer need to fear predators, but there was just something *wrong* about the structure. Muted colors flashed slowly over its flanks.

The kraken continued to move forward, using its sensory cluster to examine the structure. It seemed like the same kinds of materials that surrounded it, when it was following the air flow and feasting on the reddish-brown things with shells. That gave it comfort; that made it feel bolder. There were spaces in the structure but there wasn't room for its body. So it opened its beak and grasped the structure with its tongues and pulled hard, until a piece of the structure gave way. It pulled it back, then walked through the now open space.

It could see the prey moving. There was a large one, and a cluster of smaller ones. The smaller ones were just perfect, just the right size. Brighter colors flashed more quickly along its flanks as it felt their electromagnetic pulses and their vibrations. They felt like the first prey, the host-prey that was *so* delicious.

The prey were staying together and it wasn't ready to take the big one. It was *very* big. The kraken stayed where it was and matched the background, sending electromagnetic pulses towards the cluster of prey, hoping the pulses would confuse them. Ultimately, it needed one of the smaller prey to separate; it needed to catch one of the little ones alone.

It waited. It could feel their fear subsiding. They were starting to move independently. That was very, very good.

One of the smaller prey came towards it. It was big and pink, with very little of the thin, thread-like things growing on it. It had a flattened front on its faces, and the kraken could see sharp things in the prey's mouth that it would need to be careful with. It could see their optical sensors and the flaps that must be auditory sensors.

The prey came closer. It intensified the electromagnetic pulses. It could feel the prey's confusion; it was stumbling a bit. *So exciting!*

The prey came closer. And closer.

The kraken shot out its claws and clamped down over the mouth. The other clamping about the body just in front of the hind legs. The prey squealed! It squealed and it dug in its feet and tried to pull away, tearing at the soil. The prey squealed and released yellow fluid high in nitrogen waste from its end. It felt the warmth of the prey's body through its claws. It felt the pump moving fluid through the prey's body. It felt the lungs expanding and contracting as the prey hyperventilated. It could feel the texture of the skin, and the fat beneath the skin. Bright, bright colors flashed rapidly over its flanks as it fought to subdue its next meal.

The big prey responded to the squealing of the little one, and ran in the kraken's direction. Was this the little one's guardian? *Not good.* It did *not* want to deal with the big prey, not yet.

The kraken backed away, dragging its prize in its claws. It pushed its body back through the hole it had made in the structure, but the hole wasn't big enough for its prey. It tried to drag its prize through the hole, but it wouldn't fit. The large prey charged, and rattled the fence. *Not good.* The kraken needed to shift its grip to pull the prize through, but it didn't want to risk the prey escaping. It tightened the grip of the claw about the prey's legs until it felt the structure inside the legs snap. The prey squealed loudly from the pain. It was all *so* exciting!

It opened the claw holding the legs and reached around to grip the back of the prey. Now, with the prey's head facing away, it pulled the prey through the hole. It dragged its prize away from the opening, to get away from the larger, enraged prey. It could see the big one trying to follow, scraping at the ground, squealing as it tried to rescue the prey. The big one seemed very intent about helping the little one it was holding. *How interesting.* The kraken decided to stay where it was so the big one could watch what was about to happen and share in the experience.

It stopped the electromagnetic pulses dampening their senses. The prey became erratic and squealed even more loudly as its mind cleared. It could feel the prey's fear and pain, and the prey's squealing made the big one try all the harder to get through the opening. It began emitting augmentation pulses. The colors flashing along its flanks brightened and accelerated as it watched the big one injure itself trying to intervene. It was all so, so exciting.

It briefly considered shifting its grip again, to turn the prey and bite into its belly. It *really* wanted to feel the flesh of the belly, with its layers of fat and muscle; but it realized the orientation and shape of the prey raised another possibility, another opportunity to keep it alive even longer while it fed. And it wanted to share the experience with the prize as long as possible. The kraken locked eyes with the prey as it struggled, then opened its beak and extended its tongues to rake flesh off the legs. It could feel the warmth of the body and taste the flesh; red liquid began seeping out, and it stared into the prey's eye as it squealed in pain and fear. It kept five eyes on the prey and aimed its remaining eye at the big one as it cried in response to the prey's squeals and struggled to get through the hole.

It stretched its tongues over the prey's hindquarters and took off just a little more skin and meat. The flavors were wonderful; the taste reminded it of the host, the original prey! Then it stretched its tongues as far as it could, setting their hooks against the flesh of the prey to maximize its grip, and slowly began to pull the legs of the prey into its beak.

The prey's struggling intensified. It could feel from the prey's electromagnetic pulses that it was in pain and panicking. The pleasure was overwhelming! It wanted to bite down, but resisted the urge. It pulled the legs into its beak and partially everted its digestive organ around the feet, using its peristaltic contractions to pull the feet and the legs slowly into itself.

The sensors on its tongues were filled with the flavors of the prey. It started secreting digestive juices and it could feel the prey's feet and legs starting to dissolve. The prey screamed more loudly than ever, struggling ineffectually against the restraints of its tongues and the grip of its claws. *Oh, this is good.* This would last a long time, a *very* long time, and it would share all of the sensations with the prey.

It fed slowly and steadily, reveling in the flavors and the struggling and the squealing and the pain and the panic. It relished the frantic squealing of the big one, as it tried and tried to find a way through the hole that was just too small for it to get through. It monitored the pump it could feel in the prey's body, staring into its eye as it struggled and squealed. The little prey began to weaken and whimper; the kraken wallowed in the sensations as it felt the prey's pain and agony and terror. Finally, when the prey's abdomen was through its beak and beginning to dissolve, the piglet began to die. The pump became irregular, then stopped beating; the piglet stopped taking in air. It continued to ingest the piglet, but faster. There was no longer a need to consume it slowly. It turned its eyes from the piglet to the sow, but the sow seemed to realize the piglet was gone. It was no longer struggling to break through, but stared quietly through the hole.

As it finished, the kraken began planning. It liked this method. The piglet lasted a deliciously long time and very little of it was wasted. It wanted *all* of the prey on the other side of the hole it had made, but the big one was still there, standing guard, exhausted, staring at the kraken through the hole.

The kraken started planning how it would go about getting the other piglets, feasting on them, one after the other, until it was big enough to take the sow.

Bright colors flashed rapidly along its flanks as it thought about its future feast.

And it *really* wanted to feast on the big one.

▪ 20:52

As they assembled in the conference room, Elke stood by one of the viewing ports, arms folded, looking down at the planet's surface.

Over the years, she'd stood by such viewing ports many times, gazing down at surfaces of lifeless planets without an atmosphere and pocketed with craters, at planets with primitive life and primordial atmospheres of dense, swirling, poisonous clouds. Now, finally, a planet rich with life, with potential. She was imagining a brightly lit space elevator leading down to a beautiful city. She imagined a thriving colony, an important research outpost, a prominent economic force, harvesting rare precious metals and exciting organo-chemicals for interstellar trade. So much potential, so much at risk. Three dead. Three dead, so far.

She turned and could see how stressed everyone was. Was it only yesterday morning that she'd been summoned to Esteban's quarters?

She walked to her seat and spoke, "thank you, everyone. It's been a rough couple of days. We're making progress, but we're in a race, a race we cannot afford to lose. Let's get through our updates, decide what's next, and get some sleep."

She sensed that some of them were uncomfortable with her recommendation that they get some sleep. She looked around the room, and added, "I know you all, as I do, feel a sense of urgency. I know you're thinking there's no time for sleep. It's a crisis. I get it." She leaned forward, emphasizing her points with chops of her outstretched right hand as she looked around the room. "Use your teams. Let them work while you sleep. We're still at the beginning of this; it's been, what, 36 hours? We're still learning; we're still working through what we're facing. The time will soon come when we'll be in the thick of it, when you won't be able to pause and rest. Don't exhaust yourselves before you have to. Be smart. Be leaders. And be ready for when the situation peaks. Okay?"

She looked about the room, making eye contact with each of them.

"Okay. Dani and John, have we traced Emily's and Tam's contacts?"

Dani nodded as John responded. "As Nikko observed, we pressure test before heading to the surface. Neither Tam's nor Emily's environmental suits failed a pressure test, so we believe we can presume they were exposed on their last trip to the surface. Dani came up with lists of their intimate and casual contacts since their last participation in a landing party. We don't think we need to worry about their casual contacts, but we will be screening them. We prioritized their intimate contacts. Emily's intimate contacts are clear, supporting our theory that females cannot pass the parasites. Tam had three intimate contacts, and two were infected. Both were female, so while we will screen their contacts, we don't think they'll have passed the parasites to anyone else. Tam's intimate contacts are undergoing surgery as we speak."

"Thank you," Elke said. She opened her mouth, but paused and turned back to John. "How many people do you now have in quarantine? Six?"

"Yes, I believe that's correct." John nodded, then added. "Sick Bay capacity is preconfigured to handle the colony as it grows, including ample space, but not all of it was active. Well, it's active now. Assuming we get this under control before the kraken generate more spores, I don't think we'll need much more quarantine capacity. But I do fear that we'll need more surgical capacity and recuperative capacity, so we're bringing those resources online."

Elke nodded, "thank you, John. I should have known you'd have it under control."

She looked at George, and, with a smirk, asked, "how are your pets?"

It was just the right thing to say. The tension in the room dropped significantly. She needed them to feel confident. She feared they were going to be sorely tested, and she needed to get them ready.

George smiled and laughed lightly at the joke. "They're growing...*rapidly*. We've moved them to larger enclosures in the hangar deck labs. We tried putting two of similar size in the same enclosure, thinking we'd want to be ready to observe mating behavior when they matured. We were surprised. Almost immediately, one of them attacked the other. Interesting to watch. The attacker immobilized the other's beak and claws, used its own beak to bite off the other's sensory cluster, then held it steady while slowly consuming it, keeping it alive as long as it could. It was unnerving to watch. Unfortunately, we weren't recording, or I'd share it with you.

We've set up a variety of instruments to monitor their sensory capabilities. We know they can see fairly well, and the mobility of finger-like eye stalks provide them an excellent field of view. The part of the brain supporting their vision is well developed, suggesting they make extensive use of their eyes. We know they can hear, but brain development supporting hearing is not as extensive as vision. They can also sense vibrations, and the part of the brain supporting their vibration sensors is very well developed. Their tongues are rich in chemoreceptors and the supporting brain capacity is extremely well developed. They have olfactory sensors in their air intake, their equivalent of nostrils, but we don't think they rely much on their sense of smell given the level of brain development. And remember that mystery organ we talked about that's heavily enervated? We're now convinced it's a sensory organ, but we're not completely sure what it senses. It has no external sensory interface that we've been able to find. Based on Earth creatures, we're thinking electromagnetic, but that's just an idea at this point. We're working in shifts around the clock; I hope to have more results to share in the morning."

He leaned back and added, "I'm also getting the impression that they're very smart. I may be biased. Their ability to rapidly change their skin color and texture reminds me of Earthly cephalopods like octopus and squid. Changing skin color and texture involves very rapid visual processing of their environment and manipulation of whatever skin organs they use to match their background."

He shrugged, then continued. "Intelligence is always hard to objectively assess; you have to take into account their environment. If you were to put, say, two turtles on a tabletop, one a desert species and one a pond dweller, you'd see the pond dweller on the floor in no time, quite possibly injured. You might think the desert species was smarter. Is it? It's from an environment where a fall might be fatal, whereas the pond dweller is from an environment where it's perfectly safe to jump because you'll always land in water or, at worst, soft mud." He shrugged again, "so it's hard to assess. But we're running some scenarios to see what we can see. But there's something about them that suggests, to me, that they're…calculating."

He leaned forward again. "It's unfortunate that we can't go to the planet's surface. I'd like to know what's happening at the tsunami site. Are these things running around? Can we collect more for dissections? Can we learn anything from observing them? Does anything eat them?" He raised his eyebrows as he processed his last question. "That would be very valuable information… and very interesting to observe."

Jenna was intrigued. "You're not suggesting we bring predators up from the planet and release them to hunt the kraken, are you?"

George shook his head quickly, "no, no. Superficially, that might seem like a good idea, but I don't want to trade one infestation for another."

"I completely understand your frustration and your desire for more information," Elke affirmed. "What about drones? Did you come up with a way to use drones for observational purposes?"

He nodded, saying, "we're working on it. We hope to have a prototype to test in the morning."

Elke turned to Moira and asked, "where are we with spore filters?"

"We have a design we think will work." Moira, very proud of her inventions, gave a genuine smile. "We're testing them now and working with Nikko to put them in place. Should be complete by end of day tomorrow."

She leaned forward and flicked her wrist computer, then pointed at the display. "Here are prototype weapons that don't require as much marksmanship. The flamethrower has a backpack and a rifle-like wand; we're equipping it to be adjustable in terms of range and breadth. The shotgun shoots a spray of fine rubber pellets. Now, it's important to remember that both increase the risk of collateral damage, so even though they don't require as much marksmanship, they still need trained operators. Especially the flame-throwers, given the need to coordinate with our fire suppression systems." She flicked her wrist computer again and both vanished.

Dani nodded, "understood. We'll use them responsibly."

Jenna leaned back and folded her arms; her face had slowly shown more and more frustration as the meeting had progressed. "What about colonists?" she snapped. "Do we get armed?"

There was a long pause. Elke leaned forward, her jaw firmly set. "I thought we were past this. If you have personnel that are weapons proficient, I would like them to join Dani's security teams. If they are not weapon proficient, we should not arm them."

Jenna leaned forward in her chair and her eyes glinted with indignation. "You cannot deny the colonists their right to defend themselves. They won't stand for that…" A smug smile flashed across her face. "And neither will I."

Dani interjected, "we don't have time to train colonists on weapons safety, and I think arming inexperienced colonists may be more dangerous than our guests."

Elke stared unwaveringly at Jenna. "Let me be PERFECTLY clear." Everyone at the table snapped their heads; Elke rarely used that tone when speaking to her crew. It was a very uncomfortable noise. Cecilia and David exchanged looks. "I am responsible for the safety of *every single* colonist. I will do everything possible to keep them safe, including denying them access to weapons they cannot safely handle. If you have colonists that can demonstrate proficiency, please refer them to me or Dani. We need the help. But do NOT fight me over this. Now is *not* the time."

Jenna stared back, her face oddly blank. She unfolded her arms and looked away from Elke.

"Fine."

After a long, tense pause, Elke looked around the room. "If there's nothing else, I suggest we get some rest and resume in the morning."

She remained sitting as they stood and filed out. Alone again, she glanced out the view port and sighed heavily.

PART 4:
BATTLES

•DAY 24 05:17 PLATFORM TIME

The farmer woke up early, as was his habit. He'd grown up with dairy cows that needed morning milking and, even though he didn't expect to see cows on his farm until they were brought by future deep space vessels, he couldn't shake years of habit. Nor, frankly, did he feel a particular need to.

He dressed in a utilitarian jumpsuit, drank the last reheated remnants of yesterday's coffee, and went to do his morning's chores. He might not have cows, but the goats needed milking; also the pigs and chickens needed feeding. He'd return for breakfast, watch the news vids to catch up on current events, then head to his fields to check on his crops.

As he walked from his quarters in the agricultural ring, morning was breaking. Drum lighting was slowly brightening, just a little earlier than yesterday. The farmer was impressed with the lighting program, how it managed day lengths to simulate seasons to help crops grow. The lighting system consisted of a series of fixtures arranged in an arc high over the fields, sequenced to simulate the rising and setting of the sun.

As he approached his goat pens, he could see something was wrong. It took him a moment to realize that he couldn't hear them moving around. He was used to them anticipating his arrival.

He continued walking towards the pen. He could see over the side now, but he didn't see any goats. He could see, to the far right, a hole in the pen. How could they make a hole in the pen? He sighed. If they were roaming loose around the ring, he'd be hearing about it for weeks.

The farmer started walking around the pen towards the hole, stopped, then started again. *Should I...fix the hole...or look for the goats? And where should I...go...first?* His thinking suddenly became a little muddled; maybe he should have made a fresh pot of coffee. *Damn, I got a good amount of sleep last night...at least I think I did.* Why was he feeling so thick this morning?

He was so focused on the hole in the pen, that he hardly noticed his peripheral vision was fading. He bent down and stared at the hole; he could see the breaks were outward. Somehow the goats must have pushed outward through the pen wall. He'd never seen goats do that before. *Damn...how the hell am I gonna repair this?*

The farmer bent to look more closely, and nearly lost his balance. He was not feeling well. *What is going on?* Maybe he needed to get a checkup.

He saw a flash of motion as his chest and thighs were suddenly grasped with crushing pressure. He screamed and cursed as he struggled; the claw about his chest had trapped his left arm, but his right arm was free. He pounded at the claw, but its scaly surface was tough. He tried to tear at the claw with his fingernails, but his fingernails just kept breaking. His thighs were immobilized. He did his best to kick but could not make contact with anything.

His mind suddenly cleared. As the farmer fought, he followed the claw with his eyes; in the growing light, he could see a shape taking form. It had been perfectly camouflaged, matching the color of the grass; but now it was flashing bright colors, washing from the snout over the flanks, pulsing faster and faster. It must weigh almost as much as he did.

The claws were firm, but the crushing pressure had eased. The farmer still couldn't move, and he was certain he was badly bruised. At least he could breathe. He stared as something moved towards him from the top of the body: a limb ending in a bulge with flaps like ears and six fingers that were aimed at him. He began to realize that each finger ended in an eye: a strange, awful, alien eye. The eyes trained on him, staring into his eyes.

His shock was gradually replaced by fear as he felt the claws draw him towards it. It was pulling his legs in more closely. Suddenly, the claw about his thighs clamped down *hard* and he heard his femur break. The farmer *screamed*. The pain was unlike anything he'd ever experienced. As he screamed, he could see the colors flashing faster and more brightly, like it was reacting to his pain.

It released his legs and reached around him to grasp his abdomen from the opposite side. The other claw remained clamped about his chest. It turned him, until his feet aimed towards the creature.

It resumed pulling him slowly towards itself and, as he stared in horror, he saw the beak open and three tongues reach out for him.

The farmer screamed and swore and struggled, but the claws held him fast. The tongues, muscular and wet, stretched over his lower legs; a thousand knives shredded his shoes and clothing, ripping into the skin of his lower legs and feet. He threw his head back and shrieked. He felt warm moisture and was ashamed to realize he'd wet himself.

The tongues reached out again, stretching further, reached his knees and drew back again. The agony was blinding as they tore away little pieces of him back into the beak. He could see the eyes staring into his face. Two of its eyes continued to stare into his as the others moved up and down his body, scanning him from head to toe, then slowly returned to his eyes as the tongues tore away at his thighs. He screamed from the pain, then whimpered in exhaustion. His eyes filled with tears.

He stared at the beak and the tongues, then back at the eyes. The tongues reached out again, but instead of rasping at his legs, they held fast and pulled his ruined feet and lower legs towards the beak. The farmer could feel the strength of the smooth, muscular tongues as they stretched, one at a time, out across his body; then, as they tensed, their smooth texture was replaced with a thousand, tiny knives digging into him as they pulled him further in. He felt his feet pressed into moist muscle; waves of contractions in that muscle pulled him deeper and deeper inside the creature.

The farmer shivered with fear and panic as the eyes stared into his face. The colors flashed brighter and faster, and he felt a light burning sensation by his feet that grew hotter and stronger. Suddenly, he realized: his feet were in its *stomach*.

He struggled and struggled, but could not break free. It started to digest him and it *burned*. Oh how it *burned*!

He couldn't even scream anymore. He cried and whimpered and pleaded and prayed and struggled and stared madly back into those eyes as it slowly, steadily, drew him further and further into itself.

In an eternity of pain and agony, the farmer stared into those eyes. He felt the claws move and shift their grip. He felt the tongues leisurely moving up his abdomen to grasp his chest and shoulders. He felt the tongues slowly and steadily pull him in. He felt the burning reach his lower abdomen, until finally, thankfully, he ceased to feel anything.

▪ 05:52

Elke entered the conference room a few minutes early, holding her coffee mug. She walked to the portal and stared at the planet below. So beautiful, beyond anything she'd ever experienced or even imagined.

But three were dead on her watch, and she feared more would come. They were in a race, and she feared they were losing. They simply did not have the resources to sweep all the places the creatures could hide. And they did not yet have the means to be sure they could find creatures so adept at hiding, or to ensure they didn't reinfest swept areas.

They needed to take control of the situation, and they needed to do so today. But how?

John was the first to join her. She turned as he entered, and he walked to the portal to share the view. "How are you holding up?" he asked.

She bowed her head, then turned to look at him with a rueful smile. "About as well as could be expected."

He gave half a smile, "that bad, huh?"

She laughed, and he joined her.

"I've served with you long enough," he said, "to know you're going to take care of everyone but yourself. But we're counting on you, so you're going to have to take care of yourself as well." He put a hand on her shoulder, "I'm here for you, Elke. Just let me know what you need and when you need it."

She patted his hand, "thank you, John. I know I can count on you."

They resumed looking at the planet before Jenna walked in. She walked over, looked up at Elke, and said curtly, "I apologize for last evening."

Elke looked down at her and smiled. "Thank you. I know this is hard on all of us." She looked out the window and said, "I wish we could learn from prior experiences, but this is unprecedented. It's not even a scenario that's been played out; biochemical incompatibility was supposed to make this an impossible situation." She sighed, then looked out the window and mused, "if you want to make God laugh, tell her your plans."

As they spoke, the rest of the team arrived to start the meeting. They took their seats.

"George," Elke grasped her coffee mug as she sat. "What's the latest with your pets?"

George flicked his wrist computer, and a large enclosure appeared containing a complex array of climbing structures. He walked over to it, pointing as he explained, "we've moved them to these new, larger enclosures. Given the situation in the plenum, we wanted to see how well they climb. Unfortunately, they climb very well, jumping from one perch to another. We've also been measuring their growth rate and efficiency, and it's alarming. Their feed conversion rate puts our livestock to shame. But it's consistent with being a succession pioneer.

Anyway, we've been feeding them as much food as they'll take to determine their maximum potential growth rate. I don't think there *is* a 'maximum potential growth rate'. They're never satiated. They never turn down food; as soon as they finish one meal, they're ready for the next. They never sleep, either. If you keep putting prey in front of them, they keep eating, and you can almost see them grow." He paused, "that's a little hyperbolic, but not very." He looked at Nikko and said, "I apologize for my initially incredulous response to your data regarding the vermin catch reduction and what it meant in terms of kraken food consumption."

He spread his hands apart and the projection zoomed in on the kraken. "As you watch, bear in mind that the rat we're introducing weighs almost as much as the kraken. We've learned that the kraken are extremely strong; they can take prey substantially larger than themselves."

The kraken was walking slowly across the bottom of its enclosure, sensory cluster extended, claws folded, occasionally raising a limb to the enclosure wall. Dull oranges and yellows flowed slowly over its flanks. Their field of view did not include the top of the enclosure, so a rat seemed to drop out of nowhere. It landed half a meter in front of the kraken, froze in place, sniffing the air, then cautiously began exploring. The kraken froze and instantaneously matched the color and texture of the substrate. It partially retracted its sensory cluster close to its body.

George narrated, "we've continued to evaluate their sensory capabilities. You can't see it, but the kraken is now bombarding the rat with electromagnetic pulses that are interfering with its brain activity. What you will see is that the rat becomes sleepy and clumsy. Well, you might not notice if you don't watch a lot of rats. Let me tell you, *that* is a dopey rat.

Some of you may recall that electromagnetic fields drop off rapidly with distance. So, two things. First, the kraken is generating an impressively strong electromagnetic field. Second, as the rat gets closer, the effect intensifies considerably."

The rat moved slowly forward. The kraken began moving slowly sideways, putting itself in the path of the rat. After a few moments, as the rat moved towards the kraken, its claws flashed forward, grabbing the rat's face and abdomen. The rat struggled, kicking violently, but the kraken's claws held fast.

John overrode the projection and paused the video. "I thought you said the strike was fast enough to mortally wound prey?"

"The strike *can* be," George responded. "They appear to modulate their strikes, to avoid killing their prey outright. We'll talk more about that in a moment. For now, please note that the electromagnetic pulses interfering with the rat's brain have made it more vulnerable. It's hard to state this definitively, but I don't think the kraken would have been able to subdue the rat without the electromagnetic pulses."

George turned back to the enclosure and resumed narration: "the electromagnetic pulses from the kraken have now changed. We did some work simulating the pulses and tested our replication on some lab rats. Preliminary results suggest the pulses it emits while feeding, rather than interfering with the rat's mental acuity, actually intensifies its sensory experience."

Elke nearly spat out her sip of coffee. She looked at him with wide, horrified eyes. "Excuse me? It's doing *what*?"

George looked back at her, his face grim. "As near as we can tell, now that it's captured the rat, it has ceased interfering with the rat's mind and is now bombarding it with electromagnetic pulses that are intensifying its sensory experiences. It's going to consume the rat, slowly and selectively, to keep it alive as long as possible. And…it's using its electromagnetic pulses to…to *maximize*…the rat's experience of the pain."

They stared at the enclosure as the kraken, pulling the rat towards itself, opened its beak and bit deeply into the rat's belly. The rat screamed and struggled to no avail; it was firmly held by the kraken as it stared into the rat's eye. The kraken's tongues began rasping tissue from the rat, the waves of colors along its flanks intensifying and accelerating as the rat continued to scream.

George gestured and the image paused. "Every time they hunt," he continued, "they get better at electromagnetically interfering with their prey's brains. Every time they feed, they get better at keeping their prey alive. It used its tongues to selectively rasp tissue from the body of the rat, not damaging the heart or lungs. Kind of like their selective feeding of their host's body. The flashing of the colors isn't consistent either, it varies. We fed one of them a rat instrumented with sensors; the colors of the kraken's flanks accelerated and intensified when the rat's pain levels increased. It…" George paused, as if afraid to speak. "It…seems to sense and enjoy the rat's pain, and acts to maximize the intensity and duration of the pain."

He pointed at the rat. "This poor thing suffered for nearly an hour."

He flicked his wrist computer and the image vanished. The room was dead silent. George returned to his seat.

"To me, this is a clear indication of intelligence. There's no way they can be instinctually preprogrammed for this. They may be instinctually preprogrammed to want to keep their prey alive, but they couldn't have been preprogrammed to know the anatomy of Earth animals. They've clearly experimented and learned how to keep their prey alive as long as possible through selective consumption. They might be instinctually preprogrammed to use electromagnetic pulses while hunting, but there's no way they're preprogrammed for a rat's nervous system. They're clearly capable of experimentation to learn how to interfere with their prey's thinking and then how to augment their prey's pain and fear."

The room remained silent for a few more moments. John slowly shook his head.

After a moment, Moira spoke. "We, uh, have a spore filter. It filters down to three microns without unacceptable air flow reduction. Rather proud of that, actually; it's a conical shape that provides more surface area. Even the bigger filters for the maintenance ducts that connect the plenum spaces to the life support core are able to maintain air flow."

Nikko said, "we've begun deploying them and will continue to deploy them as Moira's team produces them."

"Any idea whether kraken could break through them?" Dani asked.

Moira paused to think, then shook her head fervently. "Shouldn't. In order to support the conical shape, we had to reinforce them with a wire frame. Why?"

Dani leaned back and raised her hands. "Our challenge has been the number of people available to conduct sweeps versus the total volume to be swept, especially given their camouflage abilities. We've had no way to ensure kraken didn't re-infest areas after we swept and moved on. If we can coordinate our sweeps with the deployment of these filters, we can declare sections of the colony kraken-free."

Elke nodded, "that's great, that's what we've needed. Can we get one of these filters to George's team to see whether a kraken can get through it?"

"Sure," Moira answered. "And, George, we'll work with you to design an enclosure. Perhaps two chambers separated by a filter, with air flowing from one to the other? We can put a prey animal in the upwind enclosure and the kraken downwind, see if it can get through?"

George nodded, "absolutely. We'll be ready."

Elke continued, "Moira, how hard would it be to come up with motion sensors? The kraken are hard to spot given their camouflage; why couldn't we use motion sensors?"

"We've talked about it…" Moira looked unimpressed. "The problem has been vermin. Sure, the vermin population is down where the kraken have been hunting, but they're not completely eliminated and they'd confuse the motion sensors, generating too many false positives. But I've been thinking about whether we can safely ignore anything below, say, a kilo, based on the growth rate of George's pets."

Nikko jumped in, "if your team can start fabricating motion sensors, we can equip the sweep teams. We can also deploy them with the filters, so we'll know if a kraken is moving near a filtered air duct."

Elke tapped her mug nervously; her mind was racing, weighing options, balancing priorities. "I know it'll take a while to ramp sensor production. Please don't delay filter deployments. We'll add sensors as they're ready, and we'll need to go back to some areas. Can't be helped. Dani, have you had a chance to investigate missing livestock and pets?"

Dani nodded. "We've verified that pets and livestock are indeed missing. Before our unwanted visitors, when we'd get such reports, we usually found that the animal had run off and either returned or was killed by a variety of accidents. It happens.

Well, since yesterday, there's been a definite increase in missing cats, small dogs, and livestock: chickens, goats, and young pigs. We've also found kraken droppings in the agricultural rings; you can spot them right away given the uric acid component. I'm convinced that we are now losing pets and livestock to kraken."

She looked around the room, and the look on her face disturbed them. "I'm concerned. Given the growth rate George has seen, and given the size of some of the missing animals, the kraken are now big enough to threaten a child. I know this won't be popular, but I think we need to consider a lockdown."

Jenna snapped, "I don't think 'won't be popular' begins to cover it."

Dani made a face. "Jenna, you have to work with us. Our understanding of the threat is evolving. It is more dangerous than we previously anticipated. Here's what I'm suggesting: we coordinate our sweeps with the installation of the spore filters. As soon as we sweep a ring segment, we can rescind the lockdown for that segment. It won't last long."

Jenna paused, then, leaning forward, shouted, "you suggest a crisis, but you REFUSE to arm the colonists! Why won't you allow them to defend themselves? They have a stake in this too, not just you scientists!"

Elke stood. She did not often use her size to command attention, but when she did, she did so effectively.

"That will be *enough*."

All conversation immediately stopped. Jenna glared at her, then turned away.

"Your attitude is deeply disappointing. I do not like repeating myself, but I will. If you have colonists that are proficient in the use of any weaponry, any at all, we will gladly equip them with whatever weapon they can safely use. Moira's team can fabricate nearly anything. We will gladly welcome their help in the sweeps. They are welcome to defend themselves and their neighbors.

What I will NOT do is hand a lethal weapon to someone that isn't trained to use it responsibly! Nor will I attempt some half-assed quick training. I will NOT have innocent bystanders wounded by friendly fire."

She leaned forward, hands on the table, and looked intently at Jenna.

"Is that sufficiently clear?"

Silence.

Still looking at Jenna, who refused to meet her eyes, Elke said, "Moira, please continue to produce the weapons we've discussed. I want everyone that's had weapons training armed at all times."

She glanced over at Dani, then continued, "we will declare a lockdown immediately. I want the sweeps organized, starting in the residential rings, with spore filters installed as you go. We can rescind the lockdowns as we progress. Nikko, I would like some of your people to do well checks in the residential and agricultural rings. Do *not* have them attempt to engage the kraken. Just do wellness checks and investigate when people report incidents."

Dani and Nikko nodded.

Elke pushed off of the table, returning to her full height. The fire in her eyes was scathing. Her hands clenched into fists at her side, a rare demonstration of her power and tenacity. It both scared and inspired them.

"I want *every single one* of these damned things caught or *killed*."

"This is going to take some time…" Dani spoke quietly.

"Then it will take time. We're done for now. That will be all."

• 07:31

The kraken moved confidently through the dark space, grasping structures and relying on its sensory capabilities to find food and avoid others like itself. Some of its kin, it could sense, had grown very, very big.

It was *hungry*. The creatures with fine gray hair and naked tails were tasty, but so small. It bombarded them with electromagnetic pulses until they could barely walk, then everted its stomach, rapidly digesting them. They were nourishing at best, but not satisfying. There was no sharing of the experience. And now it had grown to the point that they were just too small to bother with.

It desperately needed larger prey. As it moved through the dark spaces, it observed a variety of creatures in the spaces below. Some were like the original host. Tempting, but usually too large, at least for now, and they tended to move in groups. It needed to grow before it could take any of them; and to grow, it needed to feed.

It noticed that some of the creatures were not like the original prey. They moved about on four limbs, like the creatures it had hunted in the dark space. But they were clearly predatory and therefore dangerous. But it needed sustenance. It would have to start taking risks. It finally realized that these creatures were often confined in enclosures, often alone. The kraken found one and perched above it, experimenting with electromagnetic bombardments as it made loud, sharp noises. Eventually, it learned how to suppress its thinking and quiet it down.

It took that prey cautiously; its strike swift and strong, killing the prey instantly. It drew the prey up into the dark space and examined it thoroughly before ingesting it, noticing the sharp claws, sharp teeth, and powerful jaws. It enjoyed the taste, but not all the thin hairs covering it. And it missed the opportunity to share the experience of its consumption with it. Such a waste to feed on prey that way.

It found more of them, and, each time, experimented further. They were quite diverse in their size and appearance, but the kraken learned, from their consistent taste, that they were simply variations of the same prey. Soon, it was able to subdue them enough to take them alive, striking carefully to restrain their jaws and claws.

It consumed quite a few of them. They were very emotive. It was a joyous, delicious thing to share the experience of their consumption.

And so it continued to feed and to grow. Growth was essential. Growth was safety.

As it grew, it began to think about the other prey, the prey like the original host. There were so many of them, and, if they were like the original host, they would be *extremely* tasty. Most were still too large, but, as it grew, it began searching for one that was small enough to take. And alone, it had to be alone.

As the kraken moved through the dark space, senses sharpened, it detected one of them was below. It was a smaller individual, in another enclosure.

It approached, slowly and carefully.

When it was above the enclosure, it moved downward. It neared the barrier and found that, like the other enclosures, the barrier between it and the prey was merely a set of independent pieces. Pieces with gaps between them, gaps through which it was picking up delicious scents, gaps it could see through.

It lowered itself further still, extending its sensory cluster through the gaps. It peered through.

There it was! It was the same kind as the host, but *small*!

A variation? A young individual?

It was alone, and not moving; it appeared to be at rest. It had noticed that many of the prey it encountered went through these periods of quiet repose. It did not understand this state; it experienced nothing like it. But it had learned to take advantage of it, realizing that it could send electromagnetic pulses to induce it and deepen it. As it perched above the prey, it monitored the prey's electromagnetic signature, gently bombarding it in kind. It was still fine tuning its ability to deepen this prize's slumber.

It was also searching with its other sensors. It could tell there was another, slightly larger individual nearby, in another enclosure. Not moving, but not in this quiescent state. Perhaps it would be able to take that one, too.

But this one would be first. This one was perfect.

It opened its beak, extending its tongues, and gently moved the panels of the barrier aside, striving to avoid noise. It was able to open the space beneath it enough to access the prey. The prey remained in its state of repose.

It lowered itself, cautiously, slowly. *Must go slowly.*

When it was close enough, it shot out its claws, grasping the prey. Slowly, the kraken pulled itself back into the dark space, hoisting the prey along. The strike had moved some of the objects in the enclosure, making some noises. So it backed into the dark, where it felt safe, and watched as it prepared to slowly consume the prey.

Bright colors flashed along its flanks, a dizzying display in the pitch black.

The prey began to struggle, disturbed from its quiescent state. It ceased bombarding it with suppressing waves and began sending augmenting waves, experimenting as it did so. It was still trying to find the right signals.

The prey continued to struggle, to no avail, and began to make very loud noises.

It found secure footholds in the dark space and deployed its sensory cluster to look into the eyes of the prey. The prey stared back, still struggling, still making loud, shrill noises. Moisture was coming from the prey's eyes. It pulled the prey towards itself, opened its beak, and grasped the legs of the prey with its tongues. It could taste the flesh of the prey as its tongues tore through the material encasing the lower limbs. It carefully removed that material, exposing the prey's flesh. It placed its tongues along the limbs of the prey, tasting the flesh as its tongues tore into the skin, and pulled the feet into its beak.

The delightful flavors of the prey saturated its tongues. The prey was so delicious! Just like the original prey!

But, oh, this was *better. So much better!*

Before, with the host, it was one of many. It wasn't in control of the situation; it could sense the host's pain and suffering, but it couldn't manage it. It couldn't extend it or intensify it.

Now, it was alone with the prey. It was in control. It held the prey fast, learning to intensify the prey's experience. Not only was the flesh delicious; the waves of fear and pain the prey released were scrumptious. This would be *so* good; it needed to make this one last!

It continued to look into the child's eyes, as it felt the child's feet reach the inside of its digestive organ. It slowly released each tongue and stretched it over the little body, reveling in the tastes as it tore away the remaining materials encasing the prey.

It began to digest. *Slowly.*

After a moment, the prey's struggles intensified as it resumed its screaming. More moisture came from the prey's eyes as it stared into them.

It had missed this shared experience! It was so exciting, so good to feel the pain and fear coming from the prey. It loved knowing they were experiencing it together. It was eager to consume the prey and find more like it, but it needed to slow down, to make this last.

■ ■ ■

She was sitting at her work table. She was exhausted and needed to rest. But there she had work to do, and so few opportunities to get it done.

She'd forgotten how much work a toddler could be. She'd raised three other children, she laughed at how quickly she forgot. She'd read, once, that parental amnesia was not only common, but essential to human population growth. She smiled at the truth of that sentiment.

She tried so hard to focus on her work, despite her lethargy, that she barely heard the noises coming from her son's room. She sighed. *He shouldn't be awake yet; it's way too early.*

She paused, listening.

She heard more banging and wondered what could be going on. She knew the kinds of noises children made. She could usually distinguish between the sounds that needed intervention and the sounds that didn't.

Then she heard her son scream; she leapt to her feet. She ran to the door and opened it. His bed was empty, moved away from the wall. She looked about the floor, trying to figure out where he'd gone.

Then she heard him whimper. The sound came from above.

She looked up and saw that some of the ceiling panels had been moved aside. She wondered how that had happened; she wondered how her son could have climbed up there. She knew that didn't make sense, but she was foggy. She couldn't quite process *why* that didn't make sense.

She heard him whimper again. She moved forward, looking up, into the dark.

"It's all right, Tommy!" she yelled up. "Don't move. Momma will go get a light."

■ ■ ■

The kraken heard the other prey come into the room, but it wasn't concerned. It was well hidden in the dark space, out of reach. It continued to stare into the prey's eyes with four of its own, using the remaining two to monitor the other prey in the enclosure below.

It maintained its slow, steady consumption of the prey it held. The child was now deeply within it; his limbs slowly dissolving near their major joints. The prey was in agony, staring back, quietly whimpering, not responding to the noises from below.

This was exquisite. The flavors were amazing, and it could sense the prey's agony. Absolutely delicious.

But then the prey gradually stopped struggling. It seemed to be exhausted, resigned to what was happening. That was disappointing; it was beginning to degrade the experience for both of them. The kraken considered its options, and wondered whether there was another way to enhance the prey's sensations.

Then, it started thinking about the other prey. It wondered why the other prey was approaching; was it connected to the child that it was enjoying? Was it…guarding the child? Would that make it easier to capture?

It found it could simultaneously bombard the prey it was consuming, to augment its experience, while dampening the other prey's thinking. That was good; that would help it take the other prey, too. It would make it less likely to escape.

Despite the pleasure of consuming the child in its grasp, the kraken prepared for the other prey by stopping its flashing colors and matching its background.

It began to wonder whether it should consume its prey faster and prepare to take the other. It stretched a tongue across the prey's body, tasting the area above the upper limb, next to where its sensory cluster connected to the rest of the body. It wrapped the end of its tongue around the neck and head of the prey, caressing it, tasting it, rasping it, tearing it.

Yes. It was decided. As delicious as this prey was, it would need to be ready to consume the child quickly, if the other prey returned.

■ ■ ■

She left the room, still confused as to how her son had gotten into the ceiling. But she realized she needed to find a light, and quickly. She knew she had one, but she felt so, so confused. *Where did I put that light?!* She must be more tired than she realized. Maybe, when she managed to get Tommy safely back in the room, she could take a nap. She could get her work done later this evening, perhaps when his father returned.

She found the light and returned to the room. She aimed it up into the darkness, scanning for her son. As she moved the light, she saw something. She stopped and moved the light back. She could see the top of her son's head and his shoulders, but the rest of him was obscured. *Why is he upside down?!*

"Tommy?" she called out. "Tommy, what are you doing up there?"

She saw her son's head bend back, his eyes red from crying. He found her, and he called out, his voice raw and weak: "Mommy! Mommy!! Help me! Mommy!"

She shook her head; she was so confused. She watched, as he cried and shrieked that awful shriek only toddlers can make. His head suddenly disappeared, and she sensed movement above her. It was so odd. It seemed like there was movement, but she couldn't make out any shapes. She shook her head; she felt so muddled. She moved a chair to the wall where her son's bed had been, stepped up onto it, and reached up as far as she could with her light, peering into the darkness. She cursed herself for not being taller. She'd always been relatively small, and now her size kept her from being able to see up into the dark space and reach her son. *Damn it!*

Standing on the chair, on the tips of her toes, she reached up as far as she could. Suddenly, there was a flash of motion and an awful impact. Something squeezed, applying pressure around her torso and she felt herself being lifted into the air.

■ ■ ■

The guardian returned to the room, holding something bright. It was disturbing the shared experience between the kraken and its prey.

As the other prey waved the bright thing around, it decided it was time to finish this prey and prepare for the other. It extended its tongues and, as the prey tilted its head and made one last loud noise, it pulled the prey deep inside itself to finish digesting it. As its digestive organ cleared, it increased its bombardment on the other prey. It moved, slowly, cautiously, downward; this prey was bigger than the one it had just consumed. This prey was bigger than *it* was. This would be challenging!

It continued to move downward, then, when it was close enough, it shot out its claws, grasping the guardian, and pulled it upward as it backed into the dark space.

The guardian, now prey, was heavy, but not struggling, not yet. It continued to confuse the prey as it pulled it into the dark space, turned it carefully, and removed the layers of material. Tasting the guardian's delicious flesh, it could not control its excitement. Bright colors lit up the dark space.

The prey started making sounds as it began to feel the pain; still it did not resist, the kraken's electromagnetic pulses were too strong.

Finally, the kraken felt it was in a secure location. It finished preparing the prey, and stretched its tongues to begin pulling the lower limbs into itself. It pressed the guardian's legs into its digestive organ, extending its sensory cluster to stare into the prey's eyes. It ceased sensory suppression, and began to augment the prey's experiences as it began to secrete digestive juices.

The prey became alert, and made eye contact. As the prey's mind cleared, it began to understand what was happening, began to feel itself being consumed and digested. The prey began to struggle and scream.

Absolutely delicious.

▪ 08:05

Elke leaned back in the chair she kept by her communications console, another mug of coffee steaming on the counter before her. She was drinking much more coffee than usual. She could tell it was making her agitated and irritable, even as she found the sips soothing.

She'd just finished sending her latest progress report; next she would begin working through responses to her statement. *Not much help there.* It was, in a sense, reassuring that nobody could come up with better options for them; but, honestly, she would have gladly accepted some embarrassment if it came with a solution they'd missed. She just felt so inadequate, given the threat they faced.

She'd been in the Deep Space Service a very long time. Accidents happen; people get hurt. And sometimes, people die, despite your best efforts. She sighed and shook her head; she knew that she should be doing more, but wasn't sure what else to do. She felt certain that, one day, she'd look back and realize what else could have been done. She was certain she was failing her people by failing to find more solutions. But all she could do was her best. All she could do was deal with the clear and present challenges before her. She needed to lead; she needed to keep her people focused and performing at their best.

A sudden clattering startled her. She realized her hands were shaking and jiggling her coffee mug. Her anxiety level was high. The colonists were bristling at the shutdown, regardless of assurances that it would be temporary. She feared Jenna was feeding the disquiet rather than quelling it. She needed to find a way to sway Jenna. She understood her concerns about the safety of the colonists, but she'd seen too many novices injure themselves to consider handing out lethal weapons to people without training.

Perhaps it was a mistake to not continue studies of the tsunami succession. Her indecision made her stomach churn. If they knew how long it took ecosystems to recover, they might better predict how long it would take the kraken to reach breeding size. But at what incalculable risk? She'd have to ask George whether the drone surveys proved helpful in understanding how the tsunami site was progressing.

She stood and sighed. She'd feel much better once the spore filters were in place.

She looked at the time and realized their next meeting would be starting soon. She decided to take her coffee and enjoy the view from the conference room while waiting for the meeting to start. Suddenly, she heard Dani's voice.

"Governor?"

"Yes, Dani, what do you need?"

"Governor, we have a problem. We have a reliable report of a kraken attacking a colonist."

"Where are you?"

"Residential area. Ring 6, Section F, Room 34."

"On my way."

• 08:13

Jonah and Emma were in the plenum space of the first residential ring. They were following one of the security teams through the darkened space; their headlamps helped but not nearly enough. The footing was challenging in the dark, especially carrying spore filters. The filters weren't terribly heavy, but they were bulky and an odd shape. At least they'd already installed the last of the larger filters designed for the main maintenance ducts to the life support core.

They were nearly done with this ring. It had taken most of the morning, and the first segments were cleared and closed to kraken by the filters now installed in every duct. They hadn't fully appreciated just how many ducts there were. Jonah wondered about just how long this would take to finish, given the pace they were setting and the number of personnel available.

As they approached the next duct, Jonah turned to Emma. "Your turn."

She sighed, handed him all but one of her filters, and began climbing into the upper reaches of the space.

As she did so, Jonah yawned. He was suddenly, and inexplicably, tired. The work was quite challenging, and physically awkward, but not too draining. He realized he was feeling a little dopey, too; almost like he'd had a few drinks. *Odd…Should have had another coffee.*

He was looking upward as Emma climbed when he thought he saw movement. *Weird.* Nothing should be moving up here. Sometimes there would be the occasional roach or rat, but not lately. That was one good thing about the damned kraken; at least they were clearing out the vermin.

As he looked around lazily, Emma briefly lost her footing. "Ow! God dammit…"

Jonah called up, "are you okay?"

She paused, then answered. "I'm fine. Just tired, I guess. This repetitive chore is getting to me, I guess…"

He laughed. "Me, too. Must be more demanding than we'd thought. Tell you what, when we finish this section, coffees are on me."

She looked down and smirked, "deal! Just a few more. Can you see the security team? They seem to have gotten ahead of us."

He looked into the gloom. He couldn't see them anymore; they were out of range of his headlamp. He could hear them, though, in the distance. Maybe this was his chance. Emma was so pretty. Gorgeous eyes. He really liked her. Maybe, over coffee, he'd finally gather his courage and suggest getting together. *Maybe she'd come over for dinner? Ah well, once we're through this lockdown…*

"No, I can't see them. I can hear them, though." He laughed. "Hope they're not trying to sneak up on any of the kraken. They're not exactly stealthy."

He heard her laugh, then her laugh was abruptly cut off. Something about that got his attention, despite how groggy he was feeling. He peered upward. Jonah saw Emma suspended horizontally in the darkness, struggling. He could hear her gasping.

"Are you alright?"

No answer.

As he looked, he could see her moving oddly, turning in the air. Parts of her were obscured as she turned. She was staring upwards, murmuring. He began to climb. She seemed to be floating away from him, into the darkness.

"Hey!" He called out to the security team. "Hey! Can you hear me? I think Emma's in trouble!"

No response.

He continued to follow, and, as he followed, she stopped. She was suspended in the air, face up, her head towards him and her legs pointing away. Her legs seemed to end abruptly; he could not see her feet.

As he got closer, he could see, on the structures around him, shreds of her clothing. He looked again, and focused his headlamp. He started and fell backwards.

Emma was sobbing and moaning, nearly naked, random strips of her jumpsuit still on her upper body. Her lower body was completely unclothed. He was oddly, mildly aroused. He shook his head in disgust. *You horny idiot!* He rubbed his eyes and shook his head again. His mind was so, so fuzzy.

Carefully, tentatively, he made his way closer and began to make out shapes. There was something suspended over her face. It was a weird, round object, with flaps outstretched and finger-like projections, most of which were aimed at her face. Two of the fingers were aimed at him. He looked closer, and realized each ended in an eye: a weird, alien eye.

Another eye turned towards him.

A shiver ran up his spine, and he tasted something sour in his mouth. The cluster of eyes sat atop a long stalk that led back to an egg-shaped body; four stout limbs ending in feet grasping supporting structures. Bright reds and violets flashed along its flanks. The dizzying light display made Jonah nauseas.

He could see Emma's upper body was held in claws. Her legs were inserted into the splayed beak of the thing, its three tongues reaching out across her body. The tongues ripped off the remaining strips of her clothing, tearing her flesh as it did so. Blood leaked from the tears in her skin, as it pulled her into itself.

He shook his head, trying to clear the fog. A slow-building panic took hold. "Emma?!"

She heard him and leaned her head back, slowly tearing her eyes away from the thing above her. She started sobbing, "Jonah, help me...Please, Jonah, help me, HELP ME!"

Tears streamed from her eyes as she stared madly at him. Her voice grew increasingly loud and shrill. "It burns! It buuuuuuurns! Oh, God, it *burns, please*, Jonah, PLEASE HELP ME, IT BUUUUUURNS!!"

He shook himself and began screaming as loudly as he could for the security detail. He dropped all but one of the spore filters, and, in a desperate act of bravery, began beating at the claws with it.

More of the eyes shifted from Emma to him. She looked at him and continued weeping and pleading as he beat at the thing without effect. The tongues tightened around her and she shrieked as they continued to stretch out over her naked body and tear at her flesh. One of the tongues tensed and pulled her in further. Jonah lost his footing and almost fell; he dropped the filter, his only weapon, and grabbed at the supporting structures to regain his balance.

As he did so, one of the claws released Emma and snapped around his chest. The force of the blow was blinding. He was now suspended in the air, his hands and feet no longer in contact with the structure.

He pounded and tore at the claw, to no avail. He watched four of the eyes return to Emma, and she now stared back at them, shuddering and weeping as she moaned and softly whispered things he could barely hear. One claw still held her upper body, and the tongues had resumed slowly raking back her skin, destroying her now naked body as it pressed her into its mouth. Jonah could see her bare breasts and torso shredded and bleeding where the tongues had released her to improve their grip. He stopped beating at the claw, as her thighs and then her genitals and flat belly disappeared into moist, gray, pulsing muscle.

As he stared, her waist disappeared into the thing. Jonah watched Emma's back arch as her breathing became irregular, then intermittent, then stopped. As she died, the eyes shifted from her to him. The tongues pulled faster, swiftly drawing the rest of her into the mouth and the gray, moist muscle. One of the tongues reached out, caressing her head as it wrapped over her crown of blond hair. Then, without warning, it squeezed mercilessly and pulled her expressionless face inside it. The top of her head disappeared.

Emma was gone.

The other claw reached out for him, as did the tongues. Jonah could see the thing more clearly now. Bright, vivid colors continued to flash along its flanks. The tongues began tearing at his clothing as he screamed. His mind cleared as he felt his feet pulled toward the beak. A warm, moist muscle contracted, pulling him deeper into the thing. Jonah screamed, and as he stared into those monstrous eyes, he felt a terrible, horrible burning.

▪ 08:37

"Governor! I don't need you to come here. I'll meet you in the conference room."

Elke, striding purposefully through the corridors, abruptly stopped. Dani's voice had her frozen in her tracks. Anxiously, she looked up as she asked, "Dani, what's happened?"

"Best we talk in person. Out."

She stopped, taken aback by Dani's harshness. But Elke knew better than to question her. She paused to think through her best route, and then walked to the nearest lift. As she did so, she tapped her wrist computer.

"Colonial staff to the conference room. *Immediately.*"

▪ ▪ ▪

She walked into her conference room, and found Dani had beaten her there. The look in Dani's eyes, that terrible look, actually made Elke shudder. As she walked up to Dani, others trickled in and joined them.

"What's happened?"

Dani paused and looked about the room. Everyone but John was present.

"We have confirmed attacks. The kraken are attacking humans!"

George, just walking in, nearly tripped at her words. "That's not possible! They couldn't be big enough."

"I'm sorry, it's confirmed. They're huge."

Elke asked, "how big?"

"We killed one in the plenum," Dani answered. "It was over 1.5 meters, and about sixty kilos."

A pause, then, George let out a gasp as he took his seat. "That's…hard to believe." He raised his hand quickly, "I'm not doubting you! I'm just saying that's hard to believe."

"You said attacks." Elke's comment stopped everyone in her tracks. "Attacks. Plural. What else happened?"

Dani paused to collect herself. Her distraught frightened them all, especially Elke.

"We assigned a security team to sweep the plenum above the first Residential Ring. They'd nearly finished, when two of Nikko's people..." She paused and looked at Nikko, who had started out of his chair. "Two of Nikko's people were installing spore filters and fell behind. It was...Jonah and Emma." Nikko made a noise and sank slowly back in his seat. A pained expression swept across his face. Dani continued, "my security team heard screams, and backtracked. They found Jonah being consumed by the kraken. It..." She paused, looked downward and shook her head, as if trying to shake the vision from her mind.

"It was eating him. It was eating him alive. It had already digested his feet and lower legs. John's got him now."

She looked up again. "My team approached, and the thing backed away; it held him in its claws and continued feeding on him while they surrounded it. Jonah screamed the whole time. They shot it with rubber bullets, but that didn't bother it. Finally, they started hitting it with the flamethrower, carefully, so as to not hit Jonah. But it just kept backing away! They torched it until they killed it. They pulled Jonah out of its mouth and he babbled about watching it eat Emma. They took him to Sick Bay. Hopefully...John can save him."

Dani sank her head in her hands. A sickening silence filled the room. Cecilia and Jenna looked like they were going to be sick.

Elke put a hand on Dani's shoulder. "Are you ok?"

Dani looked up at her, "frankly, no."

Elke nodded, "then sit."

As Dani took her seat, Elke went to her own. The rest of the senior staff followed suit. Elke looked upward and called out, "John?"

After a pause, John's voice came over the intercom. "Yes, Governor?"

She looked downward. "How's your patient?"

"He'll live. He still has about thirty percent of his thighs; his lower legs and feet are gone. They appear to have been digested, subjected to the kinds of acids the xenobiology team determined are secreted by the kraken's digestive organ. He'll live. We'll see how much regeneration we can stimulate; whatever we can't restore, we'll replace with prosthetics. But..."

There was a pause. Elke looked up again. "But what?"

They heard John sigh. "Right now, he's in shock. He's staring and mumbling. He's endured an excruciatingly painful and terrifying experience; I can't imagine the damage to his mind based on what he's been through. Longer term? Longer term, I don't know. The electromagnetic interference from the kraken has done extensive damage to his brain. He's hemorrhaging in a number of places. Myelin sheaths have been damaged. That's what we've found so far; we're still scanning him." Another pause. "There's been a lot of advancement in regenerative therapies for brain tissue, but this kind of damage is not something we have a lot of experience with. We'll work with him, but I can't make any promises as to that part of his recovery."

Elke nodded solemnly, "I understand. I'd say do what you can, but I already know you'll do everything possible. Anything else?"

After a pause, John said, "let's not have anyone join him."

Elke paused, pained by the statement. "We'll do everything we can. Out."

She turned to George and asked, "thoughts on Jonah's injuries?"

George looked pained. "We noticed, last night, that more and more of the captive kraken have adopted this feeding strategy. At some point, as they deal with mammalian prey, they seem to discover that consuming them feet first avoids waste and, well, it lets them keep their prey alive longer. Once they try it, they stick with it. It's a horrific way to die, but it's consistent with the kraken's priorities of minimizing waste while maximizing their prey's suffering."

He sighed, "I didn't think they could manage it. The skeletal foundation for the beak is ingenious. It provides for an incredibly powerful bite, yet unhinges to facilitate ingesting large prey." He leaned forward, "overall, they're very tough and very strong. The skeletal structure, the beak, the limbs, all of it, it's just a bioengineering marvel. When this is all over, we should review their biomechanics. I suspect we'll learn a few things we can use. But, in the meantime, they're hard to kill, and now they're a threat to our people."

Elke looked at him a moment, then scoffed. "I know they're just animals, but I hate them."

George paused, then shrugged. "I completely understand. Part of me admires them as the ultimate succession pioneer, supremely adapted to their ecological niche. But they are unconscionably cruel. We can talk about this being an artifact of their parasitic phase and needing to keep their host alive. But it is, without a doubt, horrifically cruel."

Elke's eyes glazed over, a deep fire burning behind them. "Anything else?"

George nodded, "we continue to experiment to learn more about their electromagnetic capabilities. As they grow, they get more sophisticated. The largest ones can now deal with more than one prey item at a time. We've observed them suppressing two rabbits at a time; then, as they begin to feed, they can continue to suppress one while augmenting the sensory experience of the one that's being consumed. If they're getting big enough to threaten humans, we need to bear this in mind."

He paused, then said, "another thing. Remember that they're in a race to grow as big as they can and as fast as they can. If they've grown big enough to attack people…if they've grown too big to be interested in pets and live-stock…then they're going to hunt people aggressively and relentlessly."

Elke leaned back, nodding, then turned to Dani. "I presume you are now reminding your security teams to stay with the operational people accompanying them?"

Dani, still subdued, nodded.

"Daniella." Elke's tone was surprisingly crisp. "We were expecting to find smaller predators that might threaten pets and livestock, not something that would attack our colonists and friends. It's not your fault, nor is it the fault of your security team. Take corrective action and stop punishing yourself. We need you at your best. Understood?"

Dani looked up. "Yes, Governor," she spoke unenthusiastically. She turned to Nikko and whispered, "I'm so sorry."

Nikko was quiet, then said, "Dani, I agree with Elke. This wasn't your fault. I'm, well…I'm devastated, but I don't blame you. The question now is how to ensure this doesn't happen again."

Jenna pounded the table with her fist. "The way to ensure this doesn't happen again is to equip people to defend themselves!"

Elke stared at Jenna fiercely. "We will do what we need to do in order to protect the lives of the colonists. You need to decide, right now, whether you're going to be part of the team or a spectator from the brig."

"You wouldn't dare!"

"Try me."

After a long, intense pause, Jenna looked down and mumbled some unintelligible.

"We will no longer be holding staff meetings." Elke held her gaze on Jenna, but spoke to the whole table. "We will stay here until this crisis is resolved. You will direct and manage your teams from here." She looked at Moira and Nikko and pointed to a corner of the room. "Set up some comm stations over there to facilitate working with your teams. What's next?"

Dani sighed, becoming distraught again. "I believe…we may have another attack. The reason I called you earlier was that a mother and child are missing from one of the residential rings. Ring 6. Neighbors report signs of a disturbance, including panels missing from the ceiling. There's no sign of the mother and child."

Elke swore loudly, making her team jump. "What's the status of the sweeps?!"

Dani responded, "we've now cleared the first residential ring. Spore filters are in place. The residents of that ring are now free to move about as long as they confine themselves to that ring."

She paused, then continued. "I had ten teams, each team led by one of my people supplemented with one or two of Nikko's people that have some weapons training. Two teams swept each ring. Given the size of the kraken we just killed, I've instructed them to merge so there's one larger team sweeping each ring and escorting Nikko's spore filter people. Slows them down a bit, but we don't want to lose anyone else.

Ring 2 is almost clear; so far, no kraken. Rings 3, 4, and 5 are in process. In Ring 3, we found and killed two smaller kraken; one was about 15kg, and the other closer to twenty. The team of Ring 1 is now starting to sweep Ring 6. We're hoping they'll find the mother and child as they go, but it's a sweep, not a search. We must prioritize the threat to the rest of the colony."

Elke nodded. She understood.

"Have you thought about what happens *after* you've swept the residential rings?"

Dani nodded, then flicked her wrist computer. A schematic of the colonial drum appeared. She pointed at the lowermost rings, "the command rings are, for now, presumed clear. Vermin traps have remained untouched. We'll come back to sweep them when the rest of the colony is clear."

As she said this, the two bottom rings dimmed. She pointed to the next rings, "we are sweeping the residential rings from the bottom up. The first ring is clear; the second is nearly clear. We'll continue through the residential, lab, commercial, and industrial rings, though we're pretty confident that the kraken were limited to the residential rings and the agricultural rings. In fact, since we expect to find the kraken in the residential and agricultural rings, I'm considering locking people out of the lab, industrial, and commercial rings and focusing on finishing the residential sweep and then the agricultural rings."

Elke nodded and asked, "can we assume that everyone that might normally be in the lab, commercial, and industrial rings are currently locked down in the residential rings?"

"Yes. Right now, all humans are either here in the command rings, in the residential rings, or in the agricultural rings."

"Good. Clear the residential rings, then agricultural." She raised her hand and stopped Jenna, who was halfway out of her seat, preparing to object. "I *know* that leaves farmers vulnerable. We could, at least in theory, redeploy one of the residential ring teams to agriculture; but, that would slow down progress in the residential rings and the redeployed team has to try to cover a broad territory. I think we're better off prioritizing the residential rings, killing as many kraken as we can and chasing the rest. Then we will reassign all the security teams at once to methodically clear the agricultural spaces."

She swore again, then dropped her head into her hands. She was at the end of her tether. She fought back tears.

"I think that's our least bad option. We're going to lose colonists; I just don't see a way around that. But it's imperative that we find and kill every last kraken as quickly as possible. We have to kill them before they start releasing spores. We've accounted for three kraken now, right? Three are dead?"

Dani nodded, "that's right."

"And how many do we think we're dealing with?"

George worked his hands nervously. "Well, we just can't know. We can guess, based on Tam, that there are, or at least were, eighteen or twenty of them. But that's purely a guess." He turned to look at Nikko, then back, and continued. "Nikko was able to offer us some high level estimates based on the consumption of vermin. Now that we're learned they grow faster than we would have thought possible, it may be that we have a smaller number of faster growing kraken than we thought. Either because there were fewer of them than we thought, or there's been some cannibalism, which we also now know happens."

Elke stared at him deadpan. "Bottom line: you don't know."

George sighed and lowered his gaze. "Right. We don't know. We *can't* know." She didn't turn away. "Guess, then."

He paused, then said, "well…I can take a wild-assed guess if you want. I think there are fewer than eighteen or twenty. Much fewer. Why? Well, we know they're now big enough to attack livestock and people. If there were that many, I'd expect, and, mind you, this is purely a gut instinct, that we'd be seeing even more people attacked. So, if I had to guess? Ten?" He paused, then laughed nervously, "but, please, understand just how groundless that guess really is."

Elke nodded. She walked over to him and patted him on the shoulder, "I completely understand. And I agree. As awful as this…this situation has been, I think you're right. If there were more of them, we'd have more attacks. And, maybe we *do* have more attacks, and we don't know about it yet. Ultimately, we just can't know until we've found them all."

She swore, paused, then walked over to the view port in the conference room, turning her back to her senior staff briefly. The beautiful calm of Eden's atmosphere below contrasted the silence storm behind her. She paused, then addressed her staff:

"I think it's time to address the colony. Any disagreement?"

Silent nods from everyone but Jenna, who refused to meet Elke's eyes. Elke nodded, then, looking upward, commanded:

"Project this message colony wide."

She cleared her throat as she listened for the 'ping' that indicated she now had a colony wide channel. After it rang out, she crossed her arms behind her back, took a wide stance, and began:

"This is Governor Lubandi. I need everyone's attention for an important announcement. If you have been monitoring colony news, you know that two days ago, we found the remains of geochemist Esteban Kay in his quarters. You should also know, by now, that he succumbed to a parasitic infection, a parasite from Eden.

As you may be aware, this was completely unexpected due to our strict safety protocols, involving environmental suits, decontamination procedures, and the biochemical incompatibility of Earth and Eden life forms. Yet it happened, and we will post, for those interested, a brief overview of our understanding of *how* this happened.

All of this is new to all of us. This should have been impossible. We are learning as we go, and applying what we're learning to protecting our colony. Because we believe there is a risk that the parasitic infection might spread, our security teams have been sweeping the colony to find and destroy these alien organisms. It is imperative that we eliminate them before they mature and release spores that might infect other colonists." She paused, cleared her throat again, then resumed:

"This morning, we learned that these alien organisms have a phenomenal growth rate and have, over a very brief period, grown large enough to become a different kind of threat. We believe they have attacked pets and livestock. And this morning, one of them attacked and killed a member of one of our sweep teams. I repeat, one of these alien organisms has attacked and killed one of our fellow colonists. The alien was destroyed, but we know there are others, and we now know that they present a serious risk to our people.

Therefore, I am ordering a colony wide lockdown. Stay in your quarters. I repeat: for your own safety, stay in your quarters. Our security teams will continue their sweeps, and we will release residential spaces from lockdown as they are cleared. Agricultural spaces will be next. Remaining spaces, such as research, commercial, industrial, will follow. With your cooperation, we should be able to eliminate this threat quickly.

People…we are all counting on your commitment to our collective wellbeing. Our security personnel are committed to removing this threat. Your cooperation with this brief interruption of your liberties is essential to eliminating this threat to our colony, and I know we can count on each other through this trying time.

You can submit questions through the colony network. Members of our operations team will patrol residential and agricultural spaces to answer your questions and see to your needs. Thank you, and please, be safe."

She looked up again. "End transmission."

She returned to the table briefly to say, "Dani, start developing your plan for clearing the agricultural rings. Let's find them. Let's find them and *kill* them."

▪ 09:22

The goats were milked, the eggs collected, and the livestock fed, so he felt entitled to enjoy a hearty breakfast. Perhaps a little too hearty. He'd been putting on a little extra weight lately. Hard to believe, given the work, but he was getting older and he'd definitely put a little weight on as the years went by. Looking down at his plate, he patted his belly and thought he wouldn't care; but he wanted to keep his wife happy.

He was one of the older farmers in the community: a role model for stoicism, diligence, and independence. He leaned back and thought through the day. It was time to prepare the next field in the big rings. He wasn't looking forward to it, but it needed doing if he was going to be ready to seed. Worst part would be getting the kids to help; he knew they'd grumble about having to work. Besides, truth be told, with a large family, he enjoyed the solitude of his usual work schedule. But today's work needed more hands.

He drummed his right fingers on the tabletop and decided he'd take the oldest three. He'd let his wife and the younger ones sleep. He sighed, stood up, and went to pound on their doors.

■ ■ ■

The kraken walked across the open field, finally large enough to not feel fear. It had grown substantially through the night, consuming a variety of prey trapped in enclosures. It was such a wonderful experience, such amazing flavors.

But as large as it was, it needed to grow more. It no longer knew fear, but it still had an insatiable appetite. It needed more prey. It had searched all the enclosures to see what it could find. But they were all empty. It was time to move on.

As it walked, the kraken felt the presence of prey. These larger prey did not use all four limbs to walk, like the creatures it consumed during the night. The oncoming prey were familiar, in some way. As they moved towards it, it recognized the signature of the prey. They were like the original host! Oh, that was a *delicious* experience!

It needed to approach the prey cautiously; it couldn't risk startling them. The prey must not flee. It needed to get close enough to experiment with them, to learn how to catch them.

The kraken noticed a structure, a structure like the enclosures it had breached during the night to feed. That was promising. It could use that structure, match it, and observe the prey as they approached. That was fine; it had the capacity to be patient.

■ ■ ■

The farmer and his three kids made their way from the residential area of the agricultural ring to the broad expanses of the rings the drones had built before they arrived. The farmer idly scratched his left jaw and he glanced upward. The 'sky', the ceiling bearing a sequence of lights simulating time of day and seasonal day length, was much higher than anything he'd experienced for some years. He was getting used to it, but still found it a little unnerving.

They found their way using the structural supports for the 'sky' as landmarks. As they came to their storage sheds, he told the kids what he needed them to do. The oldest two went to the first shed; his daughter followed him to the second. He started tapping his wrist computer to program the equipment they'd be using that day, while his daughter opened the door.

He stared at his wrist computer, and slowly shook his head. *Huh, that's weird.* He'd reviewed the programming while the kids got dressed; now, for some reason, he couldn't make sense of what he was seeing. He could see the numbers, and he could read them, but he had no clue what they meant. *What the heck is going on?*

■ ■ ■

The prey continued to approach. It was a new kind of prey, different from what it had been feeding on. Walked on two legs instead of four. The kraken started experimenting with bombardments, applying what it had learned during the night with other prey. Some signals worked; some did not. As they approached, it kept experimenting, until it found the right signals. Now, it was just a matter of fine-tuning.

The kraken fought the urge to flash colors; it had to stay camouflaged. But it was so excited! It had them!

■ ■ ■

He was grasping his chin with his right thumb and forefinger and staring at his left wrist when he noticed his daughter staring into the shed. She wasn't moving. Standing absolutely still, she shook her head, then took a step forward. As she did so, he saw a flash of motion. As something flashed outward from the shed, she was plucked off her feet, then, with a loud exclamation, pulled into the shed.

The farmer squinted and shook his head, trying to make sense of it. He realized he didn't hear his other son and daughter. He turned and saw them standing in front of the other shed. They weren't moving. They were frozen in place, his son's arm partially raised as if he'd been interrupted in the act of pointing. The farmer grabbed his head with his hands. He was so groggy, so dizzy.

He found that it took considerable willpower to step forward. He took another step, trying to see into the dark. He heard his daughter scream.

He walked faster. At least, he thought he was walking faster. Nothing seemed to actually be happening. He needed to concentrate on moving his feet. *Concentrate, ya old bastard!*

She was screaming and calling for help. Her brother and sister did not. He took a step, exhausted. Then another step. He was consumed with the effort of simply taking a step. That, and getting to his daughter.

■ ■ ■

It grasped the prey, holding it with its claws. The prey stared back, unable to move. It did not have much of the annoying thin threads that covered last night's prey, but it did have something else, something inanimate covering its body.

It extended its sensory cluster to look closer, looking into the prey's eyes. It eased the bombardments for this one while continuing to suppress the others. As the prey's mind began to clear, it could sense the fear building. *So, so nice.*

It opened its beak as it drew the prey closer with its claws. It extended its tongues to taste the prey, tearing away the material covering its body so it could press its tongues against the flesh. It was delicious! And just like the original host! It reached out slowly, pressed its tongues against the flesh of the prey, tasting and gently tearing as it explored. Then holding fast, it pulled the prey's feet into its beak.

Yes! The flavors were so amazing, so perfect! It wanted this, it wanted *more* of this. The taste was so good, and the sensation, the growing fear and now the pain coming from the prey, it was all so delicious!

Then it noticed that the closest one was still moving, slowly, but moving. It did not want to be distracted while it shared with this prey the exquisite pleasure of its consumption. It watched the larger prey with one of its eyes, and began to understand that this individual was concerned with the little one it held. It decided to let it approach, then make it stop once it was close enough to watch.

Yes, it could feel fear and anxiety from the larger prey as it continued preparing the one it held. Oh, this would be good; this would be *delightful* for all of them.

■ ■ ■

The farmer finally neared the shed and peered into the dark. He could see it; it was huge. It was holding her, horizontally, with a pair of pincer-like claws, bright colors flashing along its flanks. A round object hovered over her body, finger-like projections moving up and down her figure. Except one, one aimed at him. One, with a weird eye, stared at him as he approached.

It reached across her body with long, muscular tongues and tore away her clothing, ripping her skin. Her legs were in its mouth up to her ankles. She stared up at the fingers, weeping and moaning. He could hear her whispering rhythmically. He realized that she was praying; it was a prayer he'd taught her when she was a little girl. It broke his heart to hear her, to know that she knew what was happening to her.

He stared. He knew he should do something. He knew he should move, but he couldn't. He felt frozen; he couldn't think. He couldn't do *anything*. He just stared as the tongues reached further and further along her now naked body, tearing away at her, leaving her bleeding. The monster's tongues pressed disgustingly against her body, tensing as they pulled her further and further into its mouth. As she disappeared inside it, her head moved slowly side to side; she stared back into its eyes and cried and moaned and prayed and pleaded.

The shears! A glint from the front corner of the shed caught the farmer's eye. His garden shears! He had sharpened them yesterday; maybe, if he could snip off that eye hovering by him, maybe it would be enough for her to escape. He screamed through gritted teeth; blood vessels popped out on his temple. *Yes, yes!* He was moving! But not fast enough. He dragged himself towards the tool; his arms and legs were absurdly heavy. He stretched out his arm, and it shook from the effort. Another meter, maybe less, and he would have them.

The eye watching him flashed back and forth between the tool and him, then, as if it understood, as if it could read his mind, the eye returned to him. His head began to hurt, worse than any headache he'd ever had before. His movement slowed. He froze, no longer able to move anything but his eyes. He stared at the shears. *So...close.* They were just beyond his reach. He shifted his eyes, tears welling, hopelessly, helplessly watching the monster as it slowly consumed his daughter.

For a terrible eternity, he watched, as her legs disappeared and her abdomen entered its mouth. He could hear her moaning intensify, the pitch of her voice heightening. He could see her head shake faster as she wept, always looking into the eyes of the thing. It slowly sucked her in, until finally, as he watched, she ceased to move, and it quickly pulled the rest of her into itself.

His beautiful, beloved daughter's lifeless face disappeared into the awful mouth of the beast. He stared and wept, unable to move. She was gone.

■ ■ ■

It walked out of the structure, looking at the one that was watching it, and prepared to take it next. Then, it looked at the other two. It paused, reflecting, then realized there was a relationship here. This one, the one that had watched it feed, this one seemed to be trying to protect the others. There was an opportunity here; an opportunity to take the other prey first and let this one, their protector, watch. They would *all* share in the experience.

It walked towards the two other prey, deciding which one to take first. As it did so, it experimented with the one that watched. It began to probe, to learn, to find ways to share the sensations with it as the feasting progressed.

■ ■ ■

As the creature walked slowly out of the shed, he could see it better. The claws had retracted, folding under it. Bright colors flashed along its flanks, making his headache worse. The odd, round thing with ears and stalked eyes was trained on him.

It walked towards him, and the farmer knew, somehow, he knew, it was coming for him. He thought of his son and daughter, frozen in place, and how awful it would be for them to see him consumed as he'd watched their sister.

Then it stopped. Two of the eyes remained on him, but the rest moved about, scanning the area around them. The eye cluster focused on his remaining son and daughter, like it had read his mind. His son and daughter, staring at it as he was, stood frozen in place. The farmer assumed they were under the monster's dizzying spell as well.

Somehow, he could sense their fear. He could feel their fear. He could see them weeping in their terror. He started to panic; the horror of watching his daughter's death was forever burned into his memory. He found himself imagining the same fate for his remaining children; as he did so, he saw the monster begin to turn towards them.

He knew. Somehow, he knew, in a way he could not explain, that the monster had decided to take his children first. That it could sense his pain as he watched. That it *liked* that pain, that it intended to take his children as he watched, consume them alive while he helplessly watched. Then, only then, would it come for him.

He *knew*.

And he could not move, he could not scream. He could not do anything to help his beautiful children.

He watched it reach his older daughter, saw her staring at it as it stretched out its claws. Slowly, gently, they stretched toward her, grasping her firmly. He watched the beak open and the tongues come flailing out. Four of the eyes hovered over his daughter, staring at her, like it was gloating over its prize. She shook her head slowly, side to side, her body shuddering as she wept. She mouthed 'no' and 'please'. One of the eyes trained on his son as he also stood helplessly, watching and weeping. The last eye fixated on him, as he watched the tongues grasp his daughter, lift her off the ground, rip away her clothing, tear at her flesh, and pull her bare, bloody feet into its mouth.

The colors along its flanks flashed bright reds and purples and oranges, faster and faster. His stomach dropped; he understood. It was *celebrating*. It was lighting up because it was *happy*. His heart ached and his soul screamed as he stood helplessly. He sobbed as it slowly, awfully, steadily stretched its tongues across his beloved daughter's naked, bleeding body and pulled her beautiful, innocent, weeping, screaming self into its mouth while she stared into its awful eyes and pleaded for deliverance.

The eye kept staring at him. No, not at him, *through* him, to his soul.

He knew. He knew it could feel his pain, his agony, his heartbreak. And it was *savoring* it. And all he could do was stare, helplessly, powerlessly stare, as it slowly consumed his daughter, and then, his son.

He could stare, and he could wait, wait for it to finally come for him.

▪ 09:53

Elke stood by the observation window, staring at the planet's surface while ignoring it completely. Colonial leaders sat at their places, used their wrist computers to work with their respective teams, occasionally consulting Elke or each other.

Moira nodded as she listened to her fabrication team lead. She beckoned to Nikko and walked to Elke. "The last set of flamethrowers are complete," she said. "They're starting on the scatterguns, and, after that, they'll work on the motion sensors."

Elke nodded, paused, and corrected her: "given what happened to Emma and Jonah, I want to prioritize motion sensors before scatterguns." She looked over them and caught Dani's eye.

As Dani walked over to them, Elke asked, "Nikko, can you arrange to distribute the remaining flamethrowers to the sweep teams?"

Nikko nodded and stepped away to instruct his team.

Dani reached them and Elke filled her in. "I've decided to prioritize motion sensors above scatterguns; I want to arm the sweep teams as best we can. We should also be leaving motion sensors in the swept plenum spaces, but I think that can wait until after the scatterguns."

"Why are we still making scatterguns?" Dani asked. "They weren't effective."

George, David, and Cecilia joined them. George said, "I agree with Dani. If they're that large, the scatterguns will be ineffective." He paused, then added, "remember that they're incredibly tough. The trilateral body design gives them three spines interconnected with rigid ribs protecting their major organs including their three brains. We haven't had time to research the interplay of those brains, so we can't even be sure whether destroying one or two of the brains would slow them down. You could try shooting the sensory cluster, but they tend to retract them into their protective sheaths, and they seem able to find prey with the pressure and electromagnetic senses that aren't part of the sensory cluster. Flamethrowers seem the most promising option."

Moira looked to Elke. "I'll cancel the scatterguns and get them working on motion sensors? Agreed?" Elke nodded, and Moira stepped away.

Cecilia asked, "What can we do?"

Elke smiled tiredly, "nothing I can think of. Just stay in communication with your people, so we know they're okay and they know what's going on."

David nodded. "Will do. But let us know what else we can do."

"Do any of you have any weapons training?" Dani asked.

David and Cecilia looked at each other and shrugged. David shook his head. George interjected, "I've used weapons while managing zoological parks: dart guns, things like that. A few of the xenobiologists have; the ones with training in whole animal specialties."

Elke tilted her head. "We'll get you armed. Give Dani the names of your team members that should also be armed." She paused, then asked, "would that suggest you have some experience tracking?"

George nodded, "Yes, most of us do. Any of the whole animal specialties: wildlife biology, ecology, like that. We've had to track animals in zoological parks, but we won't be much help in the plenum sweeps."

"But it will be *very* helpful in the agricultural rings." Elke turned to Dani, "get them armed, and be ready to bring them with you to the agricultural rings when the residential sweeps are completed."

Dani looked up at Elke. "Agreed. And I think all of your staff should be armed."

Elke nodded, "I agree. None of us are as proficient as your team, but we've all passed our annual checkouts. Still, let's prioritize getting the sweep teams outfitted with motion sensors before we arm ourselves."

- 10:19

Two security teams completed their sweep of the second subsection of Ring 4 and prepared to enter the plenum of the third. Between the physical exertion of navigating the space, and the tense anticipation of potentially engaging a life-threatening predator, they were exhausted. They took a short break near the plenum maintenance access, when a team of operations personnel reached them to deliver the next set of spore filters. They also provided four more flamethrowers so that both security teams were fully armed, and a pair of motion sensors for the two teams.

The forward security team leader explained that they'd follow the same process: head directly to the subsection's maintenance duct, get that oversized spore filter in place, then work through the remaining ducts methodically sweeping for kraken.

The operations lead tapped her wrist computer, and the maintenance access ladder popped out from the wall as the access panel opened above them. They reactivated their headlamps and began climbing. She closed the access panel behind them.

As they climbed, she realized she was more tired than she thought. She was a little older than the rest of them; and, yes, she did have a tendency to carry a few extra kilos. But she exercised regularly. She was surprised to feel such a growing sense of lethargy.

She heard a shout from the forward security team. The lead, holding a scrap of clothing, yelled back "hey! You have any idea how this might have gotten here?"

She looked up as he examined it in the light of his head lamp. "Looks torn." He paused, squinting and looked more closely. "I think there's dried blood on it!"

A voice from behind said, "Shouldn't be here, I don't like it. Stay alert, people! I don't think we're alone up here!"

"Anything on the motion sensors?"

"Not a thing."

The security team lead stopped staring at the clothing, scrunched his eyes, and shook his head. Yawning broadly, he looked down and casually asked, "what was that?"

As he did so, the motion sensors beeped. He stared quizzically at his display, forgetting how to use it. He shook his head again, his vision getting blurry. "Damn, I can't read this. Isn't that funny? I could read it a little while ago..."

Another beep. The voice from behind said, "something big just moved. Not enough to register exactly where, but close. Keep your eyes open!"

The operations lead looked about with growing disquiet. They'd been briefed that the kraken were masters of camouflage, and it was *dark* up here. She didn't like it, didn't like it one bit!

The motion sensors beeped again. Briefly.

She looked up, and, as she watched him scanning the darkness with his head lamp, still holding the strip of clothing in his hand, she saw a flash of motion as the motion sensors began beeping. This time they didn't stop.

He moved abruptly forward. "Hey!" Suddenly he was floating horizontally, his head lamp shining upward, receding rapidly through the darkness. He called out, "Hey! Help!"

Her lethargy was fading. The operations lead saw the two other members of the security team shaking their heads, one yawning as he did so, palm pressed against his forehead. Clearly they were trying to shake whatever was interfering with their thinking.

She yelled, "Get after him, dammit!"

She saw them begin to move, following him through the darkness. She could hear his exclamations fading, as was his head lamp. Then his head lamp stopped moving.

He started screaming.

The second security team and their operations support passed her, the lead saying, "wait here!"

She saw their headlamps receding into the dark. She heard voices yelling.

"I found his head lamp!"

"And there's some cloth over here, got fresh blood on it!"

"Use the motion sensor!"

They faded into the darkness as she and the rest of the operations team held their ground. She heard shouting in the distance, along with shrill screams, but she couldn't discern what was being said. She saw sudden flashes of brightness as, she assumed, flamethrowers were used.

She waited.

She began to feel lethargic again, but gradually, so gradually, that she didn't really notice. She struggled to understand what was happening in the darkness. She found that she was no longer anxious. She'd lost interest in the commotion. Somehow, it didn't seem to matter anymore. She felt good; she felt at peace. She stopped thinking about her team members lost in the darkness. She started reminiscing about pleasant memories. She yawned, and stretched, and thought about how good it would feel to rest. She yearned to lie down, someplace quiet, and enjoy these lovely memories. *Perhaps, a nap...*

She saw a sudden flash of motion, felt harsh impacts on her torso and legs, and lost her grip on the supporting structures around her. She was inexplicably floating on her back, looking up into the darkness.

As she stared upward, a strange, round object appeared over her, with odd flaps and fingers waving over her body. One by one, the fingers pointed towards her face; the large object continued moving forward, until it was hovering over her face with strange, strange eyes looking deeply into her own.

She was swept through the darkness, spinning around gently until she was moving feet first. Her head lamp fell away as she stared into those awful eyes. Her clothes were ripped away; damp muscular tongues moving obscenely over her skin. She felt their sharp edges dig into her, and her now bare feet were pressed into something moist and muscular.

Her mind cleared, and she felt the horrible burning. As she realized what was happening, she started screaming.

▪ 10:36

Dani stood slowly, staring at her wrist computer. She called to Elke, "we have a firefight, right now, in Ring 4!"

Elke stood abruptly. "Report!"

All activity in the room came to a halt.

"A sweep crew was working the plenum in the third segment of Ring 4. They found strips of cloth and investigated. My security lead from the forward team was taken. Both security teams gave chase; they eventually caught up with the kraken and destroyed it, but not in time to save the victim. While they were engaged, another kraken took a member of the operations team. Her head lamp and shredded clothing were found, but they have not yet found the kraken."

Nikko grimaced. "Who did we lose?"

Dani paused, then answered. "Anil from my team, Cheryl from yours."

Silence.

"Are they still pursuing the second kraken?" Elke asked.

"Yes, but they're not getting any readings," Dani read from her report. "They hadn't installed the spore filter in the maintenance duct yet; they think it might have fled through it."

Cecilia's voice quivered, "can it fit through there?"

"Yes." Moira responded. "The maintenance ducts are big enough for humans with tools and supplies. They go to the life support core."

"Dani," Elke nodded as she sat, "we need to secure the life support core. Where are we with sweeps?"

Dani said, "Rings 1 and 2 are clear. Ring 3 is almost clear. Ring 4, as we just discussed. Ring 5 is underway; Ring 6 has just started. No signs of the mother and child.

We are now getting other reports, and we don't have enough people to verify them. A farmer has been reported missing in the agricultural space by a neighbor that says they usually meet for morning coffee. Apparently not home, the neighbor thinks he went to feed his animals and check on his crops. Also, a young woman's neighbor reported a commotion from her quarters and that she's not responding to calls."

Elke said, "And now you have one less trained security lead. Can that crew continue with just one security lead?"

Hands outstretched, Dani said, "Not much choice." She paused, then looked at Elke seriously. "I might need to join them."

Elke nodded. "I know…it's hard losing someone from your team. I suspect we're going to lose more. We'll grieve them all when we're through this. Right now? Right now," she put her finger on the conference room table, "right now, we keep fighting." She looked at Dani, then added, gently, "right now, we need you here."

Elke continued, "remember what John said about the electromagnetic interference from the kraken and brain damage. Have the survivors report to Sick Bay for screening before they resume sweeps." She paused, then pondered aloud, "we need to continue the sweeps, but we're going to need to free up a team for search and destroy in the life support core. How much longer before we finish the sweeps in the residential rings?"

"At this rate?" Dani shook her head, "this evening, at best."

Elke nodded, then turned to Jenna and grilled her. "Why was this farmer not under lockdown?! I gave a direct order."

Jenna glared back and hissed, "because he's a free man and hasn't done anything to justify depriving his freedom. His livelihood, and our food supply, depend on his ability to do his job!"

Dani, Moira, and several other officers glared at Jenna, who stood defiantly before Elke. Elke held Jenna's stare, then spoke in a calm, quiet tone that commanded attention.

"The lockdown is for their protection. It is not imprisonment! And look what has happened when one of them failed to listen and take direction seriously."

"If you're so concerned about their protection, arm them!"

"We've been through this. Arming civilians without weapons training is a recipe for disaster. Your responses are, frankly, not rational." Elke tilted her head, considering her next move. "You can stay, or go, as you wish. You will, however, remain silent while here. Is that *understood*, Jenna?"

Jenna stood, lips pursed and nearly in tears. Fists clenched, she stormed out of the room.

Dani looked at Elke and said, "you may regret that."

Elke tapped the table with her finger. "I may...but I don't have time to think about that right now. Is there anything we can do to accelerate the residential ring sweeps?"

"Believe me," Dani responded, hands outstretched, "if I could think of anything, I'd have already done it."

Elke swore under her breath and then turned to George. "This evening, we should be ready to sweep the agricultural rings. Can you put together a team to help them hunt the kraken?"

George nodded, "absolutely."

Elke paused, "if we override the circadian programming, can we keep the lights on through the night?

"Absolutely," Nikko answered. "But I don't know what impact that will have on livestock or crops."

Elke clicked her tongue. "Probably not good, but we'll have to live with it. Worst case, we eat fabricated food from the algal vats for a while. We need to kill these things. We need to kill every last one of them."

▪ 10:43

Jenna stormed through the corridors, fuming. *Irrational? Really?!* How many are dead now, because of Lubandi's mismanagement of this crisis? And this nonsense about refusing to arm the colonists! There are alien predators hunting them, and she refuses to arm colonists!

Well, she'd see about that. Jenna ducked into a corridor, and started making calls. She would pull together a team; they would arm themselves as best they can. They'd defend their homes and their families! *And when this is over...* Jenna curled her lip. *When this is all over, I think I will speak with someone in Deep Space command. And maybe Chang. Governor Lubandi's proven her incompetence; maybe it's time for civilian leadership. I'll show them what a real leader can do!*

▪ 10:46

Nikko exchanged nods with his team, and looked up from his wrist computer "Elke, we've noticed something strange."

Elke looked up as well. "Elaborate."

"We've been monitoring the air flow for spores. Best way to do that was to monitor Eden proteins. Well, we haven't found any spores, at least not yet, thankfully. But we have started picking up some Eden proteins. And they're increasing."

George, looking concerned, asked, "can you send me some information about what you're finding?"

"Of course." Nikko flicked his wrist computer and a complex model rotated above the conference table. It was a dense mix of balls representing atoms, different colors for different elements, arranged in a complex shape.

After studying the image for a moment, George stepped back and put his chin in his hand. "Well, I can't be sure, but I think I may know what these are."

George paused, until Elke pressed, "and?"

George, with an embarrassed smile, said, "sorry. I, uh, think, or at least I fear, these may be pheromones."

David looked puzzled. "Pheromones?"

"Um yes, sorry, pheromones. On Earth, at least, a number of animals secrete pheromones. They're chemical messages that animals send to each other, for a variety of reasons. For example, bees secrete alarm pheromones when they're threatened, and other bees, picking up the pheromones, get more aggressive. This supports defending the hive."

He paused, then said, "anyway, I think these are sex pheromones. I fear the kraken are reaching breeding size, and they're starting to emit these pheromones to help them find each other in order to mate."

He turned to Dani, "you have a few dead kraken now, yes? Can I have some of my people examine the carcasses? I'd like to see if there are any changes that might be signs of sexual maturation."

"Absolutely," Dani shrugged. "We haven't moved them; they're still in the plenum spaces in Rings 1 and 2 where they were killed. They've been burned, of course, so I can't speak to the quality of the carcasses. But the spaces are clear, and your people are welcome to examine them."

George nodded, then started tapping and talking to his wrist computer, sending instructions. He paused to address Nikko: "do you have any way to filter these chemicals out of our air?"

Nikko leaned back, fingers knitted behind his head. "You're thinking we can hamper their ability to find each other to mate?" He paused, then continued, "that makes sense. I don't know; let Moira and I work on that a bit and we'll get back to you."

Elke followed the conversation silently with her eyes. Her eyelid suddenly twitched. She couldn't determine if that meant she needed more coffee, or less. "Any updates on your pets?" she asked, rubbing her eyes.

George laughed lightly, "Haha, no, I haven't checked in. I assume the team is continuing to feed them. I'll see if we've learned anything useful."

"Have you considered whether there's anything we can do to use their electromagnetic capabilities against them?"

"Against them? Say more."

Elke stopped rubbing her eye. "I don't have much more. They're using their electromagnetic capabilities to hunt, sensing their prey and subduing their prey. My first thought is whether there's anything we can do to shield ourselves from them, and my second thought is whether there's anything we can do to attack them. Is there, I don't know, some electromagnetic equivalent of a flash grenade? I'm trying to be creative."

A long, thoughtful pause cycled through the teams. Moira, having walked over as they talked, pondered aloud, "it's certainly worth exploring! We might be able to come up with some kind of shielding that protects us from the kraken's suppressive bombardments. I'm not sure about the flash grenade; that might impact the people that throw them. But definitely worth exploring." She paused, then mused, "I wonder whether we could do something with an electromagnetic pulse? That'd be easy to pull together, might really hurt them, or at least confuse them." She looked at George and asked, "can we set up some experiments in your lab?"

George nodded, "of course!"

▪ 12:08

Jenna walked through the residential rings with a growing entourage of angry colonists. They were armed with kitchen knives and clubs made from furniture they'd sacrificed in the interests of public safety. She knew she'd find kindred spirits that resented the Governor's refusal to arm colonists; when word spread of the missing children, it rapidly increased her following. Her most recent recruit, though, had been the most useful. He was a young fellow who worked in the industrial areas and had a good feeling for how things worked.

He maintained that he knew how to access the plenum, and she wanted to search the plenum of Ring 3. She reasoned that, because the attack was reported before she left the conference room, there might be abandoned weapons. It was their best shot, anyway.

He led them to a corridor access panel, manipulated the controls, and stepped back. A series of rungs popped out of the wall to form a ladder, and a ceiling panel moved out of the way. He climbed the rungs, through the ceiling and into the plenum.

Jenna and her entourage followed him, turning on the head lamps he'd provided from his workshop. It took them a while to learn to move smoothly through the space; but, eventually, they made reasonably good time.

She had only a rough idea where the firefight had been, so it took some time before they came to an area with clear scorch marks. They could smell lingering odors, fuel and cooked meats. But the carcass of the kraken had been removed. She presumed the xenobiology team had it.

They began searching the area. It didn't take very long to find two flame-throwers and four handguns. Her new friend was able to interpret the fuel gauges; he assured her the flamethrowers, though clearly used, had ample fuel.

Finally, they were armed.

Now, they would defend themselves and their homes.

▪ 12:23

The kraken worked its way through the space above the ceiling. It did not understand why, as it moved downward, it seemed to get heavier; it made it much harder to hold onto the structures, but the kraken could not deny that this was happening. It had been a while since it found any prey, and the last prey it found was pathetically small. It needed larger prey, so it needed to explore.

As it carefully crawled downward, grasping what it could with all four feet, it extended its sensory cluster. It could see a flat surface with gaps, and, through the gaps, occasional movement. There were rapidly traveling electromagnetic fields from a small number of individuals. They would make for good-sized prey.

It waited as the electromagnetic fields came and went, then realized that most were moving away. One remained nearby. The one that remained was relatively small compared to the others. That was good.

It matched the colors and textures of the space around it and crept slowly downward. It didn't want to alert the prey by breaking through the panels; perhaps it could simply strike through them. In some places, the panels were higher. Here, the panels were close to where the prey moved.

The kraken began emitting electromagnetic interference, gently, slowly. It could sense the prey was lulled into pacificity. It crept further down. The prey was horizontal, on a surface like what the original host was on. The kraken's eyes were nearly through one of the gaps in the ceiling; it could see that prey was in an enclosed space.

It could not detect any more electromagnetic fields in the area. The other prey were far away. *Now* was the opportunity. The kraken stepped up bombardment, and crept right to the ceiling panels, directly above the prey.

Suddenly, it lost its grip. The kraken fell through the ceiling panels and landed on the surface by the prey. The prey jumped up and screamed. It bombarded the prey with heavy electromagnetic impulses until the prey fell down, disoriented. It paused to place its feet, then shot out its claws. It grasped the prey with one claw around the chest and the other around the legs. The prey struggled and screamed, beating the claws with its pathetic little fists, kicking its feet back and forth. But the kraken held fast. The prey was much bigger than anything it had captured before, but it was able to maintain its grip and keep the prey under control.

It stopped sending electromagnetic pulses so the prey's mind would clear, and it extended its sensory cluster to the prey's face, staring into the prey's eyes. Bright colors flashed over its body. The prey stared back and stopped screaming for a moment. It pulled the prey closer, then opened its beak and stretched out its tongues. It ripped away the inanimate material layered over the prey and some of the flesh beneath. The prey began to struggle and screamed again. It was so delicious and exciting!

The kraken reached out with its tongues again, drew them back and pulled more flesh. It considered its prey, deciding how to proceed. It drew its tongues across the prey, raking it, tasting it, and thinking about how long it could keep the prey alive after biting through its midsection. It seemed so thin. The kraken feared it would not be able to keep this prey alive once it pierced its abdomen.

It wanted, so desperately, to keep the prey alive. It wanted to share the experience with it as long as possible. It reached out its beak and took hold of the end of one of the prey's legs. Staring into the prey's eyes, it bit down, severing the leg at the joint. It dropped the detached leg into its beak with one of its tongues. *Delicious!* The prey shuddered and shrieked.

Suddenly the kraken realized the prey was losing the precious, tasty red fluid far too quickly. It had been a mistake! The prey mustn't die too fast! It pressed the prey against the wall, opened its beak and everted its digestive organ. It used its tongues to manipulate the prey's limbs, folding them under its digestive organ. The position of the claws applied pressure to stop the loss of blood.

It stared into the prey's eyes as it screamed. Suddenly the prey stopped; it started making different, softer noises. It was a kind of slow, rhythmic vocalization. Its eyes began secreting moisture. The waves of fear and anxiety coming from the prey were so, so lovely.

The kraken pressed its organ against the prey, holding it fast and covering it completely except the eyes and its air intake. It stared into the prey's eyes and began to secrete digestive juices very slowly.

The prey resumed screaming.

The kraken digested as slowly as it could; loving the sounds and the electromagnetic sensations of fear and pain. It absorbed the prey, working from the outside, then progressively inward, making it last as long as it could.

As it digested, as it stared into the prey's eyes, it thought about the prey. There were more of them, so many more. It wanted *more*. It wanted to continue to taste them, experience them. They were marvelous. It thought about the prey's anatomy, and thought about different ways to consume them. It had to determine how to keep them alive to share the experience as long as possible.

Eventually, it felt the prey's struggling slow and stop; it felt the pump become irregular. Then, as it watched the light go out in the prey's eyes, the pump finally stopped.

It accelerated its digestion, then recovered its digestive organ. With a new plan in mind, it began searching for more prey.

▪ 13:21

Jenna led her team of angry colonists to Ring 6. There were seventeen of them now: two armed with flamethrowers, four with rubber bullet handguns, the rest with makeshift clubs, kitchen knives, and anything else they could find.

She knew the sweep teams were taking the ring segments in order, so she started her sweep with the last segment. She anticipated they'd meet in the middle. As they marched through the corridors, they made enough noise to attract the attention of colonists that came to their doors to see what was going on. An inspired few joined them.

They came to a small cluster of colonists standing in the corridor. Jenna walked towards the closest individual. "What are you all doing?"

The colonist turned his head slightly but continued to look down the corridor. "We heard a loud noise, a crash and a thud, then some screaming. We called security but they're all busy with the sweeps. We don't know *what* is going on!"

Jenna, nodding, patted him on the shoulder. "We'll take care of it!"

She turned to her followers as they cheered, declaring, "WE WILL TAKE CARE OF IT!"

Confidently, she led her team through the cluster of bystanders and down the corridor, flanked by the flamethrowers and handguns. This was how it should be: taking control of their destinies, standing up for themselves.

As they walked, Jenna began to feel tentative. She felt, confused, unsure. She *never* felt unsure! She shook her head and continued walking. She looked to her right and left; by the looks on her companions' faces, she could tell that they too were battling hesitation. She clapped them on their shoulders and said, "let's go!"

As they progressed, the weird sensation grew. She felt tired, and apprehensive. They came to a door, and, as they did so, something moved.

She stopped and stared. She could barely see it. She shook her head; she could have sworn she'd seen it move! But there was nothing there.

She took another step.

There was a flash of motion and a sudden, painful, crushing pressure about her chest and legs. She was yanked off the ground and rolled around until she was staring at the ceiling. As she floated, she watched a terrible form take shape, with bright colors flashing along its flanks. A strange round object approached her, fingers with eyes waving as they aimed at her. They stared directly into her eyes.

She could see the beak open, the tongues stretch out, and press against her legs. Before she could form a thought, her clothes were torn away and sharp hooks ripped into her flesh.

One of the eyes looked past her, at the crowd behind her. As it used its tongues to grasp her legs and pull them towards its mouth, the crowd backed slowly away. Some of the colonists screamed and ran.

Her mind was clearing. *Oh no...oh no, OH NO!*

She began to scream and yell for help. She could hear clicking sounds as the men with the flamethrowers fumbled with the controls, trying to light them. The men and women with handguns approached as well, trying to shoot the thing. But their guns wouldn't fire!

BANG!

Suddenly, there was a gunshot and a shrill voice swore. Seconds later, a loud, thumping sound echoed down the corridor. Jenna felt a tremendous flash of heat, followed by screaming and shrieking. It took all of her will power to pull her eyes away from the alien. Looking back, Jenna saw an awful ball of fire in the middle of the corridor. People were on fire! She could see them waving their arms as they burned; she could hear them screaming, slowly falling to the floor.

She began to realize it had all gone horribly wrong.

Water began falling on them as the fire suppression systems activated. She looked back as the thing pulled her into a room, then backed up onto some furniture. It moved up and over the obstacles, pulled her through falling water, and together they disappeared into the darkness.

She couldn't resist its gaze. She stared back into its eyes as it retracted its tongues, closed its beak, and carried her up and away from the light and the heat and the flames. It finally stopped. She stared into its eyes. Four of the eyes looked into her eyes; two moved up and down to scan her nearly naked, bloody body as she sobbed and trembled.

Five eyes now trained on her, with one looking past her.

It began moving again, carrying her. She tried struggling, and pounded at the claw about her chest with her fists. It slowed, and, as it slowed, she felt a wall behind her.

The sixth eye joined the other five. There was something so alien, so dark about those eyes. There was no feeling, no empathy in them at all. The beak reopened and the tongues splayed out, followed by a moist, gray muscle. It pressed forward, and the gray muscle continued to grow, pressing against her as she struggled and stared and screamed.

"No! Stop! PLEASE!!"

The claws released her as the wet muscle pressed her against the wall; it coated her naked midsection with slime, the open tears in her skin stinging as the disgusting, filthy moisture mixed with her blood. The claws loosened but she was still immobile, held fast by the muscle. She could feel its contractions as she struggled and grunted and yelled. The tongues methodically pushed against her limbs, folding them under the muscle as it continued to emerge from its mouth. Eventually, she was completely covered by it, except for her face.

The eyes looked around her and then returned, boring into her soul as she tried to turn away. Jenna cried and screamed, pleading with a monster she knew wouldn't let go.

As it stared at her, as she felt the obscene, disgusting, slimy moisture soak into every bit of her body, she began to feel a burning sensation. It started on her arms and legs, more unbearable by the moment, becoming an eternity of pain.

The monster ignored her cries and epithets. She wept, her pleadings increasingly incoherent as she felt her limbs slowly dissolve away. She begged it to just kill her.

But it wouldn't. It just stared into her eyes as she cried and screamed.

And stared.

And stared.

And savored every long moment.

▪ 13:54

Dani exclaimed in a hushed tone, "oh, my god!"

Elke turned to her, alarmed. "What's happened?"

Dani stared at her wrist computer. "There's been an explosion in Ring 6. I'm still working through the details." Everyone in the conference room stopped speaking. Any explosion on an orbiting platform was their worst fear. She shook her head, then looked up and said, "it looks like some of the colonists found a couple of the flamethrowers and one of them exploded." She resumed staring at her wrist computer.

Moira walked over as Elke asked, "is the ring okay? Do we have structural damage?"

"Shouldn't." Moira was calm at first, then her voice became more strident as she processed what she'd heard. "We designed the things so that even a fully fueled one couldn't make a big enough blast to damage the ring structure. But they have a variety of safety features! I don't see how they could possibly explode!"

"Some of my people are there." Nikko's voice got higher and angrier as he read through his team's report. "They're telling me a group of colonists came through the corridor…carrying weapons. They had stolen handguns, but none of them knew how to work them safely! Two had flamethrowers, and when one finally managed to turn his on, apparently he torched the other, *and some bystanders*, with the flame!"

Elke swore. "How the *hell* did they get weapons? Why were they trying to turn on the flamethrowers?"

"Hang on, they're still interviewing survivors. A number of colonists have been taken to Sick Bay."

Elke looked up to call John, then paused, looked down, and shook her head as she collected herself. She could only imagine the chaos in sick bay right now and did not want to add to it, but she had to ask. She looked up again. "John! John, what's happening there?!"

There was no response.

She repeated her question. After a moment, they heard, "I can't talk right now; I have burn victims coming in. I'll call when I can. Out."

Elke swore again, then turned to Nikko, "what have you learned?"

Nikko, who was talking softly to his wrist computer, looked up. "It's Mitch! I'll patch him through."

Elke nodded and Nikko flicked his wrist computer. As a smaller Mitch appeared over the conference table, Nikko said, "Mitch, you're talking to the group."

Mitch, wet from the sprinklers and surrounded by smoke, said, "Governor? It's not pretty here."

Elke leaned onto the table towards the small projection. "Report. What happened?"

Mitch sighed, looked over his shoulder, looked down, then up. "Apparently," he spoke through gritted teeth, "Jenna and some twenty or more colonists came through the corridor here in the fourth section of Ring 6. Still not completely sure who was following her or how many there were or where they got weapons. Anyway, they came across some colonists out in the hallway. They'd heard a commotion from one of the colonist's quarters and got curious. We hadn't arrived yet. They told Jenna about the commotion, so she and some of her armed people went to investigate. A large kraken grabbed her and began backing into the quarters. Most of the people with handguns weren't able to figure out how to use them. One managed to release the safety, but wasn't prepared for the kick. He accidentally shot one of the other colonists. In the arm, and with a rubber bullet, thankfully, but at very close range. Then the two with flamethrowers tried to fire. One couldn't get his to work, the other did but couldn't control the spray. He hit a few bystanders, then the other flamethrower, triggering an explosion."

He looked around, "no structural damage that we can find, but lots of incinerated materials and a lot of mess from the sprinklers. Fortunately, he didn't know how to deactivate our fire suppression systems or it would have kept burning. At some point, with sprinklers disabled, if the fire got bad enough, we'd be facing a halon dump. It could have been much, much worse… In total, we have one bystander with a gunshot wound in the arm, and seven bystanders with burns of varying severity. Two are dead: the one with the flamethrower that exploded, and the poor bastard that was standing next to him. There's no sign of Jenna."

All of the teams looked over in horrified silence. Elke dug her fingernails into the conference table. Nikko thanked Mitch and cut the signal.

Elke slowly collapsed into her chair. She put her hands together as she leaned forward, placing her elbows on the table and bowing her head slightly. "God *damn* it, Jenna…"

She turned to look at Dani. "We're losing more people to stupidity than the kraken. As you finish sweeping residential rings, can you free up a team to help Nikko's people keep the vigilantes under control?"

"Yes, of course we can," Dani nodded. "But it will slow us down when we sweep the agricultural spaces."

"Understood." Elke paused, then continued, choosing her words carefully. "I hate to say this, but with Jenna gone, we may not have as much trouble with vigilantes. But we can't risk any more of this nonsense."

She leaned back with a pained expression, her hands on the table. "Poor Jenna. Damned waste."

There was more silence, presumably in remembrance of their lost colleague. Finally, Elke broke the trance. "Alright, back to work."

▪ 14:26

George exclaimed, "we've completed the necropsies for the kraken!"

Elke turned and asked with apprehension, "Yes, and?"

"You may remember that the creatures in this trilateral super-kingdom had two modified limbs kept folded in sheaths on their backs. We thought they might be intromittent organs and internal organs associated with reproduction." George noticed a few blank looks and added, "sorry. An intromittent organ is a sexual structure, like a penis. Any land organisms that reproduce sexually will need some structures to facilitate exchanging sperm, or whatever equivalents are used by these creatures."

He paused, then continued, "well, we thought these limbs might be intromittent organs. We haven't seen any sexual dimorphism…ah, no clear distinctions between two groups that might be 'male' and 'female'. On Earth, such organisms tend to reproduce hermaphroditically. That is to say, each of them is both 'male' *and* 'female', so that any two individuals can fertilize each other."

He paused again, then said, "if that's correct, if these are sex organs and they are hermaphrodites, then, as they get bigger, we should see some changes in these organs that correspond do sexual maturity. And…that's exactly what we found. The relative size of the internal sex organs is much greater in the larger carcasses. So, we think that they are getting close to sexual maturity."

Elke stared at him a moment, then swore. "So, we don't have much time?"

George shook his head, "no, I'm afraid not."

▪ 14:33

The kraken sat above an enclosure. It was in the dark space above it, holding another prey in its claws. It had managed to catch the prey off-guard and plucked it from the enclosure below.

These creatures were absolutely delicious, the best it could remember. And it felt their pain and fear so, so richly. Consuming them was a nearly overwhelming experience.

It stared into the prey's face, watching the moisture seep from its eyes. The kraken enjoyed listening to the soft noises it made, as its mouth trembled. It languidly stretched out its tongues, one at a time, and pressed them against the prey's flesh, gently tearing the skin to maximize the release of flavors. Pulling slowly and steadily, it funneled the prey into its mouth. It felt the prey's lower limbs slowly dissolve in its digestive organ, all the while wallowing in the fear and agony coming from the prey in delicious waves.

It needed to make this last; it wanted to share this experience with the prey as long as possible.

As it was slowly feeding, it began to feel something, something different, something new. It *wanted* something, but it wasn't sure what. It felt a *need*, a *new* need. *But for what?* Then, it became gradually aware of a new smell, an enticing scent it did not recognize.

This was not prey; this was not hunger. This was definitely something new.

Distracted, it suddenly realized the prey had died. The prey's life had ended, and it had missed that moment, that wonderful, peak moment when the prey's agony and anxiety climaxed. It stretched out its tongues and pulled in the prey aggressively, finishing its meal in a hurry.

It wanted more prey. But it also wanted something else. It wasn't completely sure what that "something else" was.

But it was compelling; it couldn't be ignored. It wasn't sure what this new need was, but it knew it needed to follow that scent.

▪ 15:17

Moira worked at one end of the conference room with a few members of her team, standing next to a display and a helmet festooned with wires. They tinkered with an odd device: roughly box shaped, about a half meter square and twenty-five centimeters thick. It had a dull gray shell with mesh panels, presumably to allow air flow to cool electronics.

Moira called to George, "George, can you come here a moment, please?"

He looked up, nodded, and made his way to them. Moira handed him the device and said, "this is a prototype electromagnetic repellent. Can you, please, get this to your lab so we can see whether it has an effect on the kraken?"

He accepted the device, nodding, "I'll let you know when we're ready! Should one of your people go along to help setup?"

Moira nodded and turned to her team, "Jackie, would you please?"

▪ 16:04

George and Moira walked to the conference room table. Elke looked up from her wrist computer, "what's up?"

George answered, "Moira's team came up with a prototype kraken repellent. I had my team take it to the lab. We're ready to run a test."

Elke stood up, hopeful. "Moira, that was fast!"

As the others approached the table, Moira shrugged. "It's a pretty simple thing to construct, really. All we had to do was cobble together some standard parts."

George flicked his wrist computer and a kraken enclosure appeared. "You'll notice that we moved them to larger enclosures," he commented. "The kraken are now about 20kg each and continuing to grow rapidly." He pointed out where one of the kraken was perched, camouflaged perfectly against the background.

Jackie rolled a work cart next to the enclosure, right next to the kraken, which ignored her completely.

She flipped a toggle switch on the box and stepped away.

Initially, there was no sound. Then, they could hear the gentle whoosh of air through the vents. The activated electronics were heating up.

The kraken's camouflage slowly failed as its skin texture smoothed; its color turned to a pale gray. It turned slightly right, then left. Then, as the device warmed up, it began moving away from it.

The field of view expanded. Jackie returned to the box, lifted and held its lid with her right hand and reached into the box with her left. She raised her eyes to watch the kraken, then looked down as she manipulated settings.

Suddenly, the kraken sprinted, reaching the far end of the enclosure, raised its first pair of legs and pressed them against the clear wall of the enclosure. It moved back and forth, faster and faster, rearing up from time to time. It very clearly was trying to get away from the device's effects. Its sensory cluster was fully deployed, and the eye stalks wavered as it sought escape.

Elke smiled, for the first time in a while. "That looks promising!"

Moira explained, "it generates a strong electromagnetic field. We used some of George's data to get an idea of their sensory range. What you just saw was Jackie fine-tuning to what appears to be a particularly repulsive signal. We can't sense it, of course, but they do."

"How soon can we get some of these to the sweep teams?" Dani asked anxiously.

"The challenges are bulk and power," Moira responded. "If we had a few days, I could give you something convenient and energy efficient. For now? For now, we can put them in backpacks, but they need a great deal of power, which is hard to carry. It takes about a kilo of energy to run the thing for five to ten minutes. So, it will weigh people down pretty heavily if we're going to use the things to drive the kraken to a killing area. Once they're in the agricultural rings, we can use suspensors. That will help."

Elke asked, "if electromagnetism takes that much energy, how do the kraken do it?"

"Couple of things. When they're hunting, they're relatively close to their prey; we need to reach pretty far. Electromagnetic signals drop sharply, logarithmically, with distance. Also, from what George's research has shown, the kraken save energy through precision aiming; we don't know where the kraken will be, so we're broadcasting."

"That kind of weight would be a struggle in the plenum areas." Dani's brow furrowed. "Let's plan to equip the sweep teams after they finish in the residential rings, before they get to the agricultural rings." She paused, then continued, "can we equip the operational teams currently patrolling the agricultural rings? They're not carrying weapons right now; these things would help them protect the farmers."

"Agreed," Moira nodded. "Let's start with them. We're making the things from parts we're pulling from non-essential systems. It'll take us a few hours to get enough of these things produced and built into packs. I'll work with Nikko to get them distributed as soon as they're available."

Elke asked, "any progress with shielding to protect our people?"

Moira sighed, "not yet. We've been experimenting with helmets. We haven't come up with anything that blocks their electromagnetic signals, other than creating a countersignal to cancel them out. But the countersignal would subject the user to enough electromagnetism to cause brain damage."

Elke put her hands on her hips. Finally, they had hope, tangible solutions. But they were still far from salvation. "Please keep us posted then."

▪ 16:31

The kraken followed the scent. The scent was coming from above, so it climbed. As it climbed, it became easier to climb. The kraken felt strangely lighter.

It followed the scent, and, as it did so, it came to an opening. The scent, the amazing, tantalizing scent, flowed through, so it entered the opening and followed the scent further upward.

The darkness was complete. It continued to climb, lighting up as it did so. Soon it was nearly weightless, grasping structures as it progressed to keep from floating away. It was a disquieting feeling.

Very, very strange.

But the scent! The scent was all consuming. The scent could not be ignored.

The kraken continued to follow the scent, through the strange and dark weightlessness, until it came to another opening. The scent was pouring through.

It followed the scent. As it did so, it felt progressively heavier.

The scent led it through the dark space, then into the relatively lighter darkness. This new space was similar to the space it had been hunting. It had some anxiety. The hunting had been good there. Prey were everywhere; it fed and fed and there were still so many more to enjoy. Perhaps it was a mistake to leave so soon.

But it could not resist this scent.

The kraken continued following the scent until it came to a brightly lit space. It was like the place it came from, but the distance to the ground was greater.

It found a structure and climbed down to a broad expanse of green blades. It could see, in the distance, other structures.

There were many fascinating sensations here: new odors, strange colors. The kraken could feel the electromagnetic signals of prey. There were *lots* of prey. And that scent…

Somewhere, deep inside, it began to understand what that scent meant. It knew what it needed to do about it. It realized it needed to find prey.

It needed to find prey *now*.

■ ■ ■

The kraken followed the electromagnetic signals of the prey. It walked, matching the background as best it could, in the direction of the prey's signals.

It came to a broad, vertical barrier and walked along it, until it came to a portal. It walked through the portal into a narrower space; in this space, the upper barrier wasn't so far overhead.

It sensed prey nearby and slowed, reaching out with its senses. It targeted one of the prey, and began to bombard it with a suppression signal. The kraken felt wildly reckless; the prey was not alone. But it knew it had grown, that it had little to fear. And the scent! The scent was overwhelming. The scent filled it with need and purpose and urgency. It needed prey, and it needed prey *now*.

Do not eat. That was a strange instinct. It didn't completely understand. But the one thing it *knew* was that it needed prey *now*. It couldn't question this fact. It had to succeed.

As the kraken continued moving forward, it could see the prey approaching. There were only a few of them, and it could tell which one it was bombarding. It began to bombard the others as well, gently, just to suppress them.

It paused, waiting.

They approached. It continued its bombardments, gradually increasing the intensity.

They approached. One of them, the one it wanted, shook its head and rubbed its hand against the top of its face. The others were unsteady as they walked.

They approached. *Almost.* They were almost close enough…

It snapped out its claws and grabbed the one it wanted. It barely fought; it was so confused and lethargic. It wanted to reach out with its tongues, to strip the annoying material away that covered the delicious flesh of the prey. It wanted to *feast*, but it knew that it shouldn't.

This prey would have a different purpose.

It backed away, then turned, walking back the way it came. It began to follow the seductive scent once more.

▪ 18:11

The operations team shook their heads. They'd been walking through the residential section of the agricultural ring, doing well checks on farmers, when all of a sudden they became disoriented. The pain and dizziness intensified until they, evidently, blacked out.

Now, their minds were clearing. The team leader steadied himself against a wall. *That was the weirdest thing!* He looked around at the rest of the operations team. Everyone seemed fine. Except…

Where the hell is Bonnie?

He began to remember. He started walking, randomly, searching. As he walked, his mind cleared. He remembered. "Oh, FUCK!" He tapped his wrist computer frantically.

"Nikko? It took Bonnie! We're in the agricultural ring near the farmer's quarters. We've just been attacked by a kraken. It took Bonnie! We're going to try to follow it."

Nikko responded, "be careful! We're sending help!"

▪ 18:32

The kraken followed the scent, carrying its prey. The prey didn't move much. It was heavily suppressed, but it was undamaged. It was ready.

It followed the scent across the field. Finally, in the distance, it could see the other.

Another kraken approached, also carrying prey.

The first kraken slowed to a stop, and observed.

The second kraken was big, beautifully big, as big as it was. There was something incredibly attractive about its size and its scent. It was exquisite.

So, so lovely. The first kraken was beginning to understand. It was feeling… *aroused*…somehow. But it was a new sensation. *So confusing…*

They approached each other slowly. When they were a few meters apart, they stopped moving. They extended their sensory clusters as far as they could, staring into each other's eyes.

They felt each other's electromagnetic fields and began to match each other's signatures. They began moving to their respective left, in small, tightening circles. Their flanks flashed bright, vibrant colors, faster and faster, reaching a synchronization of urgent, brilliant gleams. Their breathing synchronized, as did their heartbeats.

As they moved about each other, the first kraken felt the strangest sensation. It felt a swelling in its back, an odd swelling. Like nothing it had ever felt before. Then it felt an odd sensation, a parting, almost a tearing. Something stirred. It felt movement, achy movement. As it felt this, it looked at the other; it could see, from the second kraken's back, a bulging, then a parting. Two limbs, draped in a sticky wetness, slowly and awkwardly unfolding from sheaths in the other's back.

And it realized that it, too, was also unfolding such limbs. It did not know it had these limbs. Somehow, as they felt each other's electromagnetic signatures, and drank each other's scents, somehow they knew what to do next.

Each released its right claw from the prey and held their prey with their left claw. They each reached out with their right claws, gently grasping the prey offered by the other. As they did so, they stopped circling.

They grasped the prey gifted to them with both claws and moved slowly forward, walking past each other, eyes staring into eyes. They stopped when they were aligned, head to tail.

They opened their beaks and stretched out their tongues; they stared into each other's eyes, then diverted two eyes to the prey gifted by their partner. They began to strip away the annoying material, tearing and tasting the delectable skin of the prey that was now just beginning to struggle and moan. Then, stretching out and firmly grasping the prey with their tongues, they each pulled the lower limbs into their mouths, pressing the prey's limbs deep into their digestive organ. They each started the peristaltic contractions that gently but firmly drew the prey's limbs deeper into themselves. As they did all this, they each stared into their prey's eyes with two of their own, and kept the other four eyes locked on their partner's eyes.

They stretched the limbs from their backs towards their partner.

The forward limb's tip opened into a cup that gently fit so perfectly, so precisely, over the tip of their partner's hind limb. Sharp spines emerged from the edges of the cup and pierced the tip of their partner's hind limb, causing both of them to shudder with waves of tremendous pleasure. It was in fact the greatest pleasure they'd ever known.

With the cup fitted over the tip of their partner's limb, they each felt, from deep within themselves, a strange, wonderful squeezing sensation. They began to ejaculate from their hind limb into the cup of their partner's forelimb, and they each opened the base of their cup to drink their partner's emissions, channeling it into the very core of their bodies. Each reveled as the fluid pulsed out of them and as they felt the warmth of their partner's fluid throbbing into them. They read each other's electromagnetic signals. They shared in their partner's ecstasy.

As this glorious experience progressed, they ceased suppressing the mental activity of the prey held by their partner; they each began augmenting the sensations of their partner's prey, sensations to maximize their partner's experience.

Both prey began to scream and struggle as they became aware of what was happening to them. They continued to stare into each other's eyes and into their prey's eyes. Slowly, they digested their prey, bit by bit, one slow centimeter after another. They stretched their tongues languidly across the bodies of their struggling victims, tasting, then tensing. They pressed their tongues against the prey's flesh, tearing and savoring, always pulling the prey steadily into their digestive organs. They each delighted in the sounds and agony of their prey.

They continued to exchange fluids laced with sex cells, slowly ingesting the squirming, suffering prey gifted them by their partner. Sharing this meal together, joining with each other and their prey, the entire scenario was nearly overwhelming.

It was the most intense and satisfying pleasure of their lives.

■ ■ ■

The operations team, still somewhat confused, relied on memory fragments. Collectively, they pieced together what must have happened. As they remembered, they started following the corridor in the direction the kraken had taken Bonnie.

They came to the bulkhead connecting to the broad, immense agricultural ring. As they passed through, they could hear, in the distance, faint screaming.

They moved cautiously, knowing they were not armed. The team lead tapped his wrist computer: "Nikko, come in. We're in the agricultural rings; I've sent you our location. We think we hear Bonnie in the distance, and we see tracks we presume are from the kraken. Are you able to see anything?"

Nikko responded, "We've dispatched a security team that just finished a sweep. They're on their way, but it'll take some time to reach you."

The team leader ground his teeth in frustration. He had to make a choice. The screaming continued in the distance. He stomped his foot and swore under his breath.

"We're approaching carefully. Send the sweep team after us. Out." They continued walking in the direction of the screaming.

It was awful, truly awful. He'd heard screams of pain and panic before. He'd been in the deep space service for decades, and he had seen many terrible things But he'd never heard anything like this before.

He wanted to run, to help, but he had orders. They had warned him how dangerous these things had become. It was in his team's best interest to be careful, to hang back. But those *sounds*!

As they approached, they began to make out shapes. He saw the shapes of two kraken, standing side by side. They were each eating someone alive. He saw Bonnie. She was *alive*.

He tapped his wrist computer rapidly. "Nikko? Nikko! Are you seeing this?! We need help NOW!"

▪ 18:41

Nikko flicked his wrist computer and the two kraken appeared over the conference table. They were each two meters in length and probably weighed over 150kg. They were now significantly larger than the average adult human.

He muted the volume. The senior staff and their teams stared at the projection, barely comprehending what they were watching. Elke asked through gritted teeth, "how far away is your security team?!"

Dani responded urgently, "any time now!"

George leaned in and stared. "They're mating. See? See the limbs on their backs? They're connected with each other and they are exchanging gametes, or whatever these things use for sperm."

David exclaimed, *"why are they eating those poor bastards?!"*

George hesitated, then offered, "some predators, especially cannibalistic predators on Earth, bring prey as a gift. As long as your partner is busy eating, it's not thinking about eating you." He straightened and grimaced. "We should expect other kraken to exhibit this behavior."

They stared helplessly as the two kraken continued to slowly ingest their victims. Even with the volume muted, they could see the victims pounding weakly at the claws that held them. They could see their facial expressions as they screamed in pain.

Elke slowly pounded the table with her fist as she stared at the image.

"They should be close." Dani held up her hands about a meter apart, then slowly moved them together. As they did so, the image became smaller as the field of view expanded. They could see the security team approaching. There were five of them, two carrying flamethrowers and three with rifles. They were running. As they approached the kraken, they switched to a brisk walk.

When they were within ten meters, the leader, carrying a flamethrower, began yelling. The kraken ignored him. He nodded to his partner and both moved their flamethrower wands. Flames were now burning brightly, projecting out of the wands about twenty centimeters.

The two of them, followed by the rifle carriers, split to surround the kraken. The lead, now two meters from the flank of one of the kraken, worked the controls again, and the flame shot out, touching the flank of the kraken. It completely ignored the attack, except to divert an eye to watch them. Its mate did the same.

Now both flamethrowers were engaged. Their operators waved them side to side, slowly, over the flanks of the kraken. The kraken closest to the team lead grasped the farmer, now ingested up to her hips, and, as the farmer screamed, pulled her the rest of the way into its gullet. The senior staff watched her disappear, arms trailing, until the kraken snapped its beak shut. Its partner did the same with its victim a moment later.

Their back limbs disengaged, folding back into their sheaths. Both monsters turned, and aimed all their eyes on their tormentors. The flamethrowers left black, smoking scorch marks as the kraken turned to them.

Suddenly, the flamethrowers stopped moving. The riflemen stopped moving. The kraken moved to the side, evading the flamethrowers as they continued to burn. The security lead grimaced, but did not move.

The kraken walked slowly towards their tormentors, eyes wavering as they examined the team and their weapons. Each opened its beak, extending tongues. They wrapped their tongues around the flamethrower wands and pulled. As they did so, the flames shifted. The kraken sidestepped with surprising speed and grace, given their size. They continued to tear away the wands and rifles until, one after the other, they relieved the security team of their weapons, dropping them to the ground. The flamethrower wands scorched the grass black as they continued burning.

The kraken slowly reached out their claws and each grasped one of the security team leads. Using their tongues to strip them, the kraken tore away skin and flesh and pulled the leaders into their mouths.

The leaders began struggling, while their support teams remained motionless. They began to pound on the claws grasping them, and called for help. They screamed in pain, while the kraken stared emotionlessly into their faces.

As the leaders were slowly consumed, the burns on the kraken's flanks slowly, visibly healed.

■ ■ ■

Elke tore her eyes away from the unfolding horror. "Daniella! I want the rest of your team, equipped with flamethrowers and electromagnetic disruptors, in there NOW!"

Moira tapped away at her wrist computer and leaned towards Nikko. "Nikko's people are carrying the first four of the disrupters to the agricultural rings. They'll meet your security people there. They're also carrying fresh flamethrowers."

Dani nodded as she spoke to her wrist computer. Sweat beaded on her forehead. "I'm sending four more of my people with supporting personnel from Nikko's team. The remaining two are finishing the last residential ring sweep. we must finish that, or we risk these things reinfesting the residential areas."

She turned to Elke, consumed with emotions. Those were *her* people out there; *her* people were under attack. She should be there; she *wanted* to be there. Without thinking, her hands flexing, she backed away and edged towards the door. "I should *go*, I should be *out there!*"

Elke stopped her. "NO!" Dani threw a look her way, then lowered her head. Elke said, "I need you *here*! Trust your teams to handle it! Just make it happen *faster*, dammit!"

She turned to George and asked, "now what?"

He looked up at her. "I'm...not completely sure what you're asking. Evidently, the kraken have continued to develop their electromagnetic capabilities. I don't know how many they can suppress at a time, but I do know there must be a limit. The disruptors should help."

He paused, then added, "if you mean what happens next with the kraken, I don't know. But we need to kill those two before their mating results in the release of spores. But I have no idea what that will look like or how long before it happens."

Elke nodded grimly. "Thank you."

She sighed, swore under her breath, and pounded her fists into the table. She looked on with terror and rage as the two kraken slowly ingested the screaming security personnel. They slowly walked back into position, unfolded the limbs from their backs, and resumed their coupling.

The room was tense as they waited, staring at the image projected over the table. Suddenly, the kraken disengaged. Their flanks stopped flashing. They paused their consumption of their victims. From the room, they could see the security team advance. Four carried flamethrowers. They were supported by four of Nikko's people, each wearing a bulky, heavy backpack, each tapping away at their wrist computers as they approached the kraken.

As they advanced, one kraken stretched out its tongues, grasped its struggling prey, and abruptly pulled the poor, screaming man into itself. His screams were cut short as the beak closed. The other kraken, after a pause, followed suit. The two of them began backing away.

The rifle-bearing members of the first security team shook their heads, then slowly advanced. They picked up the dropped scatterguns, and began firing at the kraken. The four flamethrowers advanced aggressively, walking past the riflemen, and began spraying their flames across the bodies of the kraken.

The kraken backed away faster and faster. The flamethrowers pursued rapidly to surround them and cover them with fire.

Consumed by flames, the first kraken, then the second, stopped moving.

George, after a moment, yelled, "Daniella! When they're dead, it's important that they stop! We need to examine the bodies; we have to understand what happens after they mate. We need to know what was changing inside them."

After a pause, Elke nodded. "He's right." Dani frowned, hesitated, then spoke into her wrist computer. A moment later, the flamethrowers relented.

"I'll send my team," George said.

■ ■ ■

Elke paced back and forth. "I want a *methodical* search and destroy plan for the remaining kraken in the agricultural spaces."

"Understood," Dani nodded. "Our teams just finished sweeping the residential rings."

She stood and flicked her wrist computer. A model of the agricultural rings appeared over the table. Pointing, she said, "this is the agricultural ring built by the deep space probes before our arrival. It's 500 meters wide. We distributed soil and planted grasses when we arrived, what, three weeks ago?" She shook her head in disbelief, and continued:

"Livestock are kept in large pens until the grasses are well established. There are storage and utility buildings, and there are structural members supporting the 30 meter high ceilings. These structures bear lighting banks that simulate the relative movement of the sun and seasonal day length variations. This ring is adjacent to the agricultural rings that we brought with us as part of the *Lucky Strike*. They contain the living quarters of our farmers, as well as the hydroponic farms that sustained us during our journey, and finally, the now empty enclosures that held our livestock until our arrival.

We know that a number of farmers are missing, presumably taken by the kraken. We know we've lost livestock, but we don't know how many. The 500 meter wide ring represents a little over 785,000 square meters of surface area that must be swept. Then we must go through the rings we brought, which have a lot less surface area, but those rings have plenum space and unused storage areas that may harbor kraken."

She flicked her wrist computer as she took her seat. As the ring image dissolved, she continued, "we plan to have the surviving security personnel, and their supporting operations people, armed with flamethrowers in a line across the ring. They will be accompanied by additional people with electromagnetic disruptors, who will, in turn, be followed by suspensor pallets with spare energy packs. They will advance around the ring, flushing the kraken with the disruptors, and destroying them with flamethrowers. They will continue until no more kraken are flushed. Then they will move to the rings from the *Lucky Strike* and use the model that we applied in the residential rings."

Elke, no longer pacing, took her seat. "How long will that take?"

"I don't know how many times they'll need to walk around the ring to clear it. I would guess most of the day."

Elke nodded, then asked for comments.

"Is it possible," George asked, "to filter out the pheromones? It might suppress the mating behavior."

"We're working on it," Moira interjected. "Let me get an update."

George nodded, then cleared his throat. "We, uh, talked about arming some of the xenobiologists that have weapons experience. Do you want them to join that line?"

"We could use the help," Dani said.

George nodded, "let them know when to be there."

Elke turned to Moira, "when will you have weapons for them, and for the senior staff?"

"They're ready!" Moira said. "I'll have my team bring them here. Nikko, can you lend me a few hands?"

Nikko nodded.

Elke rubbed her temple, "I suspect we'll see more kraken mating. George, can we presume they'll be attacking farmers?"

He nodded sadly.

She turned to Dani and asked, "is there anything we can do to protect them? Instead of a lockdown, can we gather them someplace and guard them?"

Dani paused to consider the situation. "Maybe? There should be some empty livestock pens in the rings transferred from the *Lucky Strike*. We could hold them there. But that would require the security teams to escort them. I'm not sure it's worth it; we might be better off just clearing the agricultural space."

Elke paused and rubbed her temple again, then closed her eyes and pinched the bridge of her nose. *Are there any good options? Any choices that don't involve preventable deaths? At least my officers made a choice when they signed up; what about my colonists? What right do I have to decide which of their lives to risk? What is the least bad option?*

She nodded, then stood and looked about the room. "Okay! We will prioritize clearing the space. But keep your team on alert, and be ready to protect anyone that's being attacked." She paused, then asked, "what can we do to monitor the agricultural rings and reduce our response time?"

"It's a big space to monitor," Moira stretched out her hands in exasperation. "The broad ring has high ceilings; we could fly drones in there. We should be able to configure them to find the electromagnetic signatures of the kraken."

Elke nodded, "do it."

▪ 21:48

Moira flicked her wrist computer. A hovering drone provided a projection of the security and operations personnel forming a line. Once the line was ready, they advanced. Another drone, equipped with an electromagnetic field sensor, moved ahead of them, scanning for kraken. They walked through the young grass, stepped around storage buildings, and navigated around empty livestock pens.

"Shouldn't there be livestock in those pens?" Elke's eyes widened with the realization.

Dani responded in disbelief. "I would...presume so? Wait...look there. There's a gap in the pen. Damn it, the kraken must have been in there. It's impossible to know whether any livestock escaped."

Elke looked at her incredulously. "Do you have ANY idea how much livestock we've lost? Does ANYONE know??"

Dani shook her head, "no, not yet. We'll need to assess once we're through this."

The line continued until it came to orderly rows of young crops. The team lead shouted something, and they carefully advanced to minimize disruption. Their initial tension ebbed as the line continued to move forward. Tired, they became distracted. They began to have side conversations.

As they watched, some of Nikko's operations people arrived carrying flamethrowers and scatterguns for the senior staff. Nikko told them to put them on a worktable in the corner.

The line progressed. All were tired and discipline lagged. Some of them were falling a bit behind. Elke noted weapons were not being held firmly. Dani stepped back and snapped some orders through her wrist computer. The leads got the message and firmed the line.

The drone suddenly accelerated and hovered over a structure next to some crops. The line broke and formed a broad circle around the shed, then moved slowly inward. A voice was heard from the line: "I think we have one. I'm feeling a little fuzzy, but there's too many of us, it can't dampen effectively."

When the circle was twenty meters wide, the kraken moved and began flashing colors along its flanks. It was nearly two meters in length. It moved forward with its sensory cluster held close. One of the leads raised his arm and shouted. The line advanced, surrounded their quarry, and waved their flamethrowers over the kraken as it desperately sought an escape. George reminded them that he needed to be able to examine the remains. Dani spoke into her wrist computer and the flamethrowers ceased.

George watched two of his xenobiologists step forward cautiously towards the smoking husk of the kraken. It did not move, and they began cutting into its flanks, reaching into the body. The senior staff watched as George's crew performed their necropsy. Cutting through the kraken's tough skeletal framework was challenging, but they were able to pull the dorsal segment away and examine the reproductive organs beneath it. Hasani's voice rang out as he conducted his examination:

"These organs are not...how we've seen them before. I want to get them back to the lab when we're done, but I'm guessing this individual has mated. This organ seems to be the one that produces the sex cells that are transmitted between them. It's larger than we've seen. And..." he paused as he moved to the other organ, "this one seems to be the organ that receives the sex cells from the partner. This one is also much larger than we've seen."

He continued to work with it. "This is all preliminary, of course. I need to get this back to the lab, but I'm guessing that these are very early diploid offspring. They're about two centimeters in length. They're connected by a cord to the wall of the organ, so I'm thinking it's something akin to an Earth mammal's placental arrangement." He paused, then added, "we haven't had any opportunities to examine any gravid Eden animals, so I have no idea what I'm seeing. But it sure *looks* like that to me! The individuals are, at this point, essentially thin and wormlike; it's hard to guess what they would have developed into. I'm guessing they didn't take long to get this big, and we should expect them to be growing as rapidly as the kraken. We should be expecting gravid kraken to be hunting even more aggressively to support the young they are carrying."

"Hasani's right," George spoke fervently. "If they carry the diploid generation internally, and if they are feeding them through something analogous to a mammalian placenta, then yes, I expect the gravid kraken to be *ravenous*. Remember: at every step along the way, their evolutionary strategy has been to grow as rapidly as possible."

Elke stared downward, at nothing in particular. Her hands hung by her side. She straightened herself, and turned to Dani. "Should we rethink moving the farmers?"

"I don't see how we can guard them, and still hunt down the remaining kraken!" Dani responded. "We just don't have the people..." She pulled back her shoulders, "and what we just heard is very concerning. We know nothing about the diploid generation, but if that's the generation that produces spores, then we may not have much time to eradicate the kraken. If we fail, we're going to have spores spreading through the colony. Yes, some farmers may get attacked...*will* get attacked. But..." She paused, and looked around at all of the frightened faces listening in.

"But if we don't kill every last kraken? We're all dead."

▪ 23:19

The gravid kraken moved boldly through the field, unbothered and without camouflage. It had grown large enough. It knew the others would be more interested in mating with it than eating it, so it no longer feared for its survival. It flashed bold colors along its flanks as it hunted; it no longer cared about remaining unseen, unless it was to ambush prey.

It had mated twice already. It enjoyed mating, the sharing of pleasure and the slow, exquisite consumption of exchanged prey. The kraken could feel the young moving within its body. They were growing rapidly. They needed sustenance, so it needed prey.

It had found more of the structures containing four legged prey and feasted upon them. Most were small and, though tasty, not worth keeping alive long enough to share the experience. It suppressed their thinking, breached their enclosures, and rapidly consumed them.

Some of the prey were larger. They had very little of the annoying threads on their flesh, and their taste was reminiscent of the host. The kraken breached their enclosures, suppressed them as it consumed their young, then consumed the mature individuals more leisurely, sharing the experience with them.

But as the young within its body grew, as they drained its body for sustenance, the kraken felt it could no longer satisfy the demands of its young with smaller prey, nor could it take the time to really enjoy larger prey. It just needed to feed. The kraken knew that if it could not capture and consume enough prey, the young within its body would begin to consume each other. So it roamed, reaching out with its electromagnetic senses, and searched for the signatures of larger prey.

▪ DAY 25 02:27
PLATFORM TIME

As the security line advanced, they came across more deserted livestock pens. Every pen was compromised; every pen was empty. It had been some time since they had seen any livestock. The wildlife biologists would halt the line periodically as they encountered signs of predatory activity. They noted prints, suggesting large kraken had broken into the pens. Droppings suggested that they had fed on the livestock. Occasionally, they would examine the droppings and find artifacts indicative of the kraken's other prey: pet collars, shreds of clothing, wrist computers, jewelry, etc.

Drones patrolled in front of them. When they detected an electromagnetic signature, they raised an alert, and the line would encircle and dispatch the kraken. The wildlife biologists examined the carcasses. They'd confirmed, in the lab, that the young they found within the kraken's bodies were, indeed, diploid. But, being so early in their development, it remained unclear what they would become. They were, like the first, still small and wormlike. Support personnel took the samples and headed to the labs for further analyses.

Then they encountered a particularly large kraken as it moved through a field. It was three meters in length and over 200kg, the largest they'd encountered. They surrounded it and, with their electromagnetic disruptors and flamethrowers, subdued and killed it. When they examined its reproductive organ, they found the diploid young were larger and more developed. They sent those young directly to the command center.

■ 03:41

When his team arrived with the kraken's reproductive organ in a large sample chamber on a suspension pallet, George led them to a corner of the conference room as the others followed. The chamber was equipped with gloves and sleeves that allowed him to work with the organ within the isolation chamber. He reached through the sleeves and gloves, selected a scalpel from the tool kit in the chamber, and extended the incision made in the field until he'd butterflied the organ. He laid the two halves side by side.

As he sifted through the unborn offspring, he observed, "I'm counting dozens of individuals. They are more developed than the ones Hasani described in the field. They range from about eight to twelve centimeters in length."

He held one in his left hand and it began to squirm.

David made a gagging noise. "It's *alive*?!"

George looked up and nodded. "I thought they might be." He returned his attention to the offspring in his hand. Using his right hand to follow a cord emanating from its belly he continued to observe:

"This cord seems to be like a mammalian umbilical cord. Except that…" He trailed off, clearly caught off guard. "That's interesting; this cord branches from cords going to…" He continued sifting through the now slowly squirming offspring with his right hand. "…these individuals. Let's see if that's an anomaly." He put the one in his left hand down and used both hands to sort through the unborn creatures. "Yes, it's consistent. I'll have my team do a detailed dissection, but it looks like they're in clusters of up to ten individuals sharing a cord to an attachment point. We'll need more studies to confirm, but I'm thinking that each implanted fertilized cell develops into a set of genetically identical offspring. And there are some attachment points with no young. Either they've failed to thrive and were reabsorbed, or…"

He picked up one of the larger ones with his left hand and began to slice its belly with the scalpel held in his right. As he did so, it squirmed more violently. He continued to slice, despite the struggles of the unborn creature in his hand. As he did so, parts of a smaller individual spilled onto the floor of the chamber.

"I thought so! They're cannibalizing within the parent's body. Happens with some live-bearing species on Earth, some species of sharks for example. The mother's body accepts implantation from a large number of offspring;

if she can't consume enough prey to keep them all developing, they consume each other until they reach a number her hunting success rate can support. Once again, remember the goal is to produce the greatest possible number of spores for the next opportunity. It's a brutally efficient reproductive strategy."

Elke made a face, "doesn't endear them to me, though."

George laughed and said, "nor me."

The unborn creature continued squirming after the incision, then slowly ceased to move. He put it down, and it remained motionless. The others in the reproductive organ continued squirming. He put the scalpel down, then pulled another out and, using both hands, did his best to examine it.

"It's clearly the triploid body design we've seen in the other generations. I'm seeing six limbs but, right now, they're buds, not differentiated into legs and claws. Two short tails like the kraken generation." He probed at the dorsal surface. "No sign of the reproductive structures we saw in the kraken generation, no surprise there, as this generation wouldn't be mating. I'd guess further examination would find some small, vestigial limbs."

He continued to examine the body, now also squirming. "The body is longer and thinner than the kraken. I can feel the trilateral skeleton. It feels tough, but this is an unborn individual. No idea how close this one is to birth."

He paused and looked up from the unborn young. "You know, this is going to sound stupid, but I just realized we have not seen anything in these reproductive organs connecting them to the outside of the kraken's body. I wonder how these young are born?" He shrugged and resumed his examination. "We'll want to figure that out. Meanwhile, let's look at our little friend's head." He tried a few manipulations of the dorsal surface and, eventually, a limb everted from the forward sheath similar to the sensory cluster of the kraken. George grunted, "not fully developed. But that's not a surprise. For now, at least, it's a similar sensory cluster on a short limb."

He shifted his attention to the front of the creature. It had a beak similar to the kraken generation, but it seemed longer, relative to the body. As he worked with it, the beak suddenly, violently, swung to the side, opened and then shut down on George's finger. He swore, but held onto the creature.

"Hah! That hurts, actually. The beak can't compromise the integrity of the gloves, but our little friend is applying a surprising level of pressure to my fingertip!" He used his free hand to work with the beak. Eventually, he managed to pry one of the three sections up enough to release his finger.

"Well, that was stupid." He laughed, then continued to work with the creature. Eventually, he put it back into the reproductive organ. As he did so, he peered into the organ, and pulled another one out. Its beak was open and long; tentacle-like tongues were wrapped around the body of one of its siblings. It was actively pulling its smaller sibling, headfirst, into its mouth.

"Pleasant little critters, aren't they?"

He returned the cannibalistic predator and prey to the reproductive organ. He selected another, larger individual, and, while holding it in his left hand, reached for a work board with his right. He placed the creature on the board, then inserted large pins through parts of its body to immobilize it against the board.

"Normally, I would not consider it acceptable to dissect a living specimen. But first, I don't know how to humanely dispatch this thing. Second, I have no idea how to anesthetize it. And third, we don't have much time. So, I hope that nobody here will judge me."

Elke patted him on the shoulder, "do what you have to do."

He nodded, and began to work with the scalpel. The creature squirmed under the pins. First, he opened the torso of the creature and carefully eviscerated it; then he made skillful incisions around the head to examine the structure. It ceased to struggle.

"Okay, first thing, we do see a similar digestive arrangement to the kraken. Therefore, I expect these things to be active hunters like their parents. I also see what looks like the electromagnetic organ, so I suspect we'll see that capability as well. I don't see the reproductive organs like the kraken generation. That's not surprising."

There was a long pause. "I do see a relatively large organ back here; not sure what this is." He continued to probe it. "It isn't particularly enervated, so I don't think it's a sensory organ. It isn't connected to the digestive tract or the respiratory system. It has a lot of circulatory investment, though. And..." he paused, looking more closely. "It does seem to have a pathway out of the creature's body. We'll need to watch these to see what it is."

He paused. "I wonder. That new organ...I wonder if that's a sporangium." He looked at the others and their confused expressions. "Ah! Sporangium. That's an Earthly organism structure. Essentially, it's where the spores are produced and held until they are released. I wonder; everything about it is consistent with being a sporangium equivalent. We'll see..."

He moved to the head. "Okay, similar to the kraken but the head is relatively bigger. The beak, as I mentioned, is very powerful. I see…Nine tongues. Interesting; they seem relatively longer, compared to the body, than the kraken generation's three tongues." He laughed to himself. "Even more like the kraken of legend, then." He leaned back, pulled his arms out of the gloves, and turned to the others.

"Well, now we have a preliminary idea of what we'll be facing if we don't kill all the kraken. The clusters are interesting…You know, all of the kraken we're dealing with are from Esteban, so they're all genetically identical. But let's imagine we're on the planet's surface, at the tsunami succession site…"

He shrugged and tilted his head to the side a moment, then continued. "We would have a number of individuals emerging from parasitized hosts. All of the individuals from one host would be identical, but individuals from different hosts would *not* be identical. They'd emerge, consume their host if that's an option, then start hunting. We know they're willing and eager to consume each other. We can also presume that there are predators that arrive during succession to either compete with them or prey on them.

So, when the kraken reach sexual maturity, they're likely to encounter partners that are not genetically identical to them. They mate and exchange gametes, the Eden equivalent of sperm." He sat down, then continued. "So, are kraken promiscuous? If they're not, if they only mate once, then we have two haploid individuals mixing their sex cells. They're not diploid, so all the sex cells from an individual are the same, so all the offspring within a given kraken would be genetically identical.

But, if they *are* promiscuous, if they have multiple mates, then they'd have some genetic diversity in the fertilized eggs that implant. Each fertilized egg would divide to form a cohort of genetically identical siblings that are not genetically identical to the other cohorts sharing the womb.

So, when the mother's feeding stalls, they consume each other within the womb until they reach a population sustainable by the mother's hunting prowess. We didn't find any individuals in that womb that weren't sharing an attachment point, so, it seems that, when they consume each other, one cohort attacks and consumes all the members of another cohort. That might suggest that, once they're born, the members of a cohort work together."

He paused, then said, "ultimately, they're born. Somehow. Possibly by eating their way out of their mother. There are organisms on Earth that do this. Mites, as I recall, their offspring eat their way out of the mother and consume her body before seeking other food."

He looked up and around. "Of course, we'll need to continue to do our best to kill all the kraken. If we don't, we should expect one or more of them to birth dozens of these diploid offspring. I don't expect them to have long lives. In our admittedly limited sampling during the short time since we arrived, we haven't found any diploid individuals running around. This suggests a short life focused on developing and releasing spores.

Beyond that, we can't say much. I'm not sure how big they are at birth, how fast they grow, how they behave. But the branched placentas make me wonder whether the members of a cohort work together."

He sighed heavily. "We're making progress, but we have a long way to go."

▪ 04:33

It didn't sense potential prey. There were some smaller creatures in the area, but they were no longer worth the effort. It could sense, in the distance, the prey it enjoyed the most: the same species as the host. But they were moving in a large group, and there was something about their movement that suggested danger. The gravid kraken wasn't hungry enough to risk that, at least not now.

It remembered that it had success hunting prey in small numbers from above; it needed to return to the dark spaces from which it had been hunting. It could feel its young restlessly moving inside its body. It could feel their desperate hunger.

It needed to feed, and soon.

The kraken found a structure and began to climb, eventually reaching the bright objects high above. It searched and found a way through into the dark beyond. It began to roam, searching for prey, trying to remember the way back to where it had successfully eaten.

It felt a commotion within itself. The young were hungry, *very* hungry. Soon, they would reach the point of consuming each other. The kraken didn't have much time.

It found its way back to where it had hunted before, where the dark space extended further down. It began to sense the electromagnetic signatures of prey. It made its way down, then across, until it was just above four of the prey moving about an enclosure. Two were larger and two were smaller, presumably younger.

It suppressed them. It had found, before, that suppressing the prey gradually worked best. But it did not have the time. The kraken increased the suppression aggressively. They all stopped moving.

It opened its beak and used its tongues to set aside the panels beneath it; when the opening was wide enough, the kraken lowered itself into the enclosure. The four were there, frozen in place. It could sense that they could see it; it could sense their fear. It could feel the delightful electromagnetic pulses of their anxiety.

It could also tell that the sudden, intense suppression damaged the smallest. That prey twitched and, as it watched, the prey fell to the floor and spasmed violently. It decided to take this one first, before the damage compromised the prey's ability to experience pain. The kraken extended its claws and grabbed the small, spasming prey as the others looked on. It wanted *so* badly to take its time and enjoy this one; the small ones were often quite emotive, generating delightful waves of fear and pain. But its young were so restless inside its body. It desperately needed nourishment.

The kraken could sense the fear and the anxiety from the others as it held the tiny prey, and it decided that enjoying those sensations would have to do. Until it could satisfy the nutritional demands of its young, it could not fully enjoy its feasts.

It opened its beak and extended its tongues, tearing away the material encasing the prey. It grasped the lower limbs and began to pull the prey inside itself, digesting the prey far more quickly than it would have liked. The kraken could sense the prey's terror and agony; it felt the tiny prey shudder and heard it making sharp, shrill sounds as its was consumed. As the prey was absorbed and the nutrients flowed through its body, the kraken could feel its young slow their movements. Their hunger was abated.

It released the prey from its claws, extended its tongues across the face and head, and pulled the prey deeply into its digestive organ. Savoring its final cries, the kraken reached for the other small one. It held that one and used its tongues to strip it of its annoying material casings as it futilely resisted. It also consumed this small one too quickly, consoling itself with the pleasure of its flavors and the waves of anxiety and pain coming from the larger ones.

Continuing to steadily digest the second small prey, it began to wonder whether the small ones were the young of the larger ones. It watched them with two of its eyes, observing their stares as it consumed their young. Wondering whether they had a parental connection, the kraken thought about how it might feel to see its young consumed and realized that was part of the fear and pain it was sensing from the larger ones. It began to understand why their pain and fear was so delightful; it would have to remember this for when it had more time to enjoy its prey.

As the small prey expired and it pulled the last of it into its digestive organ, it realized its young had calmed. That was good. That meant it could take the time to really enjoy the larger two. It turned to them, looked into their eyes, and decided to take the larger one first.

• 07:12

After consuming the last of the prey, the gravid kraken returned to the dark and continued hunting. It didn't take long for its young to drain its blood-stream of available nutrients. It became increasingly reckless as the young, and their hunger, both grew. As it moved through the dark, it found more prey clustered in enclosures: sometimes individually, sometimes in small groups. It consumed individuals rapidly without taking the time to truly savor the experience; with small groups, after consuming one or two, the hunger of its young was satiated long enough to indulge in a more leisurely feeding process.

The young continued to grow, making its body heavier and heavier. It affected the gravid kraken's balance as the movements of the young began to change its center of gravity.

Suddenly, strangely, it stopped feeling hungry. As long as it could remember, it had *always* been hungry. But now? Now, it wasn't hungry anymore.

Now, it needed to find a safe place. It needed to *hide*.

The gravid kraken climbed higher into the dark space, to the levels where its body seemed less heavy. It searched for the wide portal it had traveled through, following the enticing scent of its mate. After a time, it realized that it felt air movement and that movement suggested the portal. It followed the air movement until it reached a barrier that had not been there before. It worked on the barrier, using its beak and tongues and claws, finally managing to tear through it. It entered the portal and continued to ascend, getting lighter; it shifted its efforts from supporting its weight, to maintaining contact and not bouncing away.

It searched for someplace safe to rest.

- 07:54

Elke found herself staring out the window again, holding a mug of coffee. How many times had she refilled this mug over the last couple of days? How long since it had been washed? She shook her head, sighed, and sipped. The heat was soothing, and the caffeine, necessary.

The line of security and operations personnel continued to advance around the agricultural ring, but it had been some time since the drone last detected a kraken. The last one they'd destroyed had been in the process of consuming a pen of pigs. It, like the other kraken they'd found, was gravid with diploid young. It was the fourteenth kraken they'd destroyed. They had no way of knowing how many had escaped Esteban's quarters, or how many might have been lost to cannibalism or accidents. But, based on what they'd found in Tam's quarters, it was a number consistent with their expectations. Still, the stakes were incredibly high. Absolutely no diploid young could be left alive; if just one lived long enough to generate and release spores, they were finished. So, they continued to advance while the drone flew one full rotation.

As they sat around the conference table, Elke took stock. Everyone was exhausted. It had been less than twenty-four hours since she'd declared a crisis. It felt like it had been much, much longer. She walked to the table and, as she sat down, looked upward wearily.

"John? John, can you join us?"

They heard his voice, "I can join by voice only; I shouldn't leave Sick Bay."

"What's your situation there?" she asked.

After a pause, he responded. "We have six patients, all rescued from kraken attacks. They've each lost their feet and, to varying degrees, their legs. We're doing what we can to stimulate tissue regeneration; I'm feeling optimistic that we'll be able to restore their legs and feet. Time will tell. They each also suffered some degree of brain damage from the electromagnetic bombardments they endured. I'm similarly optimistic that they will recover physiologically, but I do expect some memory loss, cognitive damage, and post-traumatic stress disorder that will take time to treat. We'll also need to monitor them for long term damage; extended exposure to intense electromagnetic fields can cause cancers. Nothing we can't handle, as long as we catch it early."

He sighed, then continued. "We also have seven burn victims from the flamethrower explosion. Two of them lost limbs; we are considerably less able to stimulate regeneration when limbs are lost to burns. Three more have severe burns over a significant percentage of their bodies, and the remaining two have localized burns. We are doing what we can, but I'm not optimistic about the chances for the severe burn cases. The two with localized burns should be okay. Beyond that, I still have some patients under observation after removing parasitic worms. We have collateral injuries while chasing kraken in the plenum: fractures, lacerations, and so forth. We weren't ready for this." The pain in his voice was obvious. "The team is performing exceptionally well, but they're tired. Someday, when this is a thriving colony, our successors will, unfortunately, have incidents that result in large quantities of concurrent casualties. But it wasn't supposed to happen when there were only a thousand of us. We're stretched thin, and we're hoping we turn the corner soon."

Elke nodded and fingered her coffee mug, trying to not become emotional. "We are hoping the same, John. You're welcome to listen to the rest of the discussion if you want to understand where we are. Or, at least, where we *think* we are." She paused and looked around the room. "It looks like we've destroyed them all, but how can we be *sure*?"

George looked exhausted. He rubbed his face as he sighed, "honestly, we can't. We haven't picked up one of their electromagnetic signals in some time, but there could be kraken that are simply in places inaccessible to our drones." He turned to Nikko and asked, "are you still picking up pheromones?"

"No," Nikko responded. As he spoke, Elke couldn't help but notice how disheveled everyone looked. Everyone was tired; she wondered how she looked to them. "The levels have been dropping since we began clearing the agricultural rings. We haven't detected any pheromones in a few hours. That's good, isn't it?"

"It *should* be a good thing," George shrugged. "It could also mean that any remaining individuals are no longer seeking mates, and so they're no longer generating pheromones to attract each other. We don't know enough about their lifecycle to hazard a guess. Certainly, continued pheromones would tell us we still had a problem. But the absence of pheromones? Not conclusive."

Elke again played with her coffee mug; she felt her anxiety rising in her stomach. "What are you learning from your kraken necropsies?"

George sighed, and threw his hands up. "It's damned peculiar! Every kraken we've examined was gravid. The young were in various stages of development, but none were much further along than the ones we examined here; it's limiting our ability to understand what the diploid young will be when they're fully developed.

But here's what was strange. You may remember that the womb we dissected earlier..." He pointed his thumb over his shoulder at the isolation chamber still in the corner, "that one had a clutch of forty-six individuals, with clusters of six to ten individuals sharing a cord and attachment point. Some attachment points were completely barren. We believe that a dozen fertilized eggs were implanted, then divided into these clusters; some of the clusters devoured other clusters when the parent couldn't maintain a high enough predatory success rate.

Others had a dozen or so fertilized eggs, but the eggs did not divide, so we found larger individuals. Again, there was prenatal cannibalism. And, then we found some with a smaller number of much larger individuals in smaller clusters, maybe three or four in a cohort. So, where does that leave us? Our current working theory is that each kraken is placing an evolutionary bet. At some point in the early development of the diploid young, some environmental signal, a hormone from the parent or something about the consistency of the food supply in the parent's blood stream, whatever it is, is triggering different developmental strategies in the eggs. That could lead to anything from a few robust diploid offspring, to packs of smaller, presumably cooperating diploid offspring.

Once again, it's an amazing evolutionary strategy to deal with a volatile environment. Larger disruptions may take longer to stabilize than smaller disruptions, and these things are able to respond to all sorts of environmental scenarios. Or it's driven by something about the parent's hunting experience. The individuals with a few large offspring were, forgive me, the ones that were hunting colonists. The individuals with larger numbers of smaller offspring were found in the agricultural rings and were feeding on our livestock." He shook his head, then said, "we know, for Earth mammals, environmental factors can affect the development of unborn offspring, like stress hormones. So it makes sense something like that could happen here. Who knows? Maybe the parent is able to make choices about what kinds of reproductive strategy they think will be most successful. Maybe there's some degree to which the parent is able to customize their young."

He sighed, then, with an embarrassed smile, added, "if they weren't killing us, I'd admire them."

He stopped smiling and became very serious. "We still haven't figured out how they give birth their young; there's no birth canal that we can find. This is pure speculation on my part, but there are some Earth organisms whose offspring eat their way out of their mother's body. Or perhaps they're never actually born; perhaps they produce and release spores from within the parent's body. I doubt that, but it's possible. In either case, we should consider the possibility and search for kraken carcasses. I'd guess they'd look for a sheltered location before, well, birthing their young.

On a slight side note, we've slowed the feeding rate for our captive kraken. We don't want to starve them, but we don't want them breeding, either. And, of course, we're keeping them separated. I believe we should keep them alive until we're through this crisis, in case they're of some research value. Once we're confident we're clear and safe, we should sacrifice the captive kraken and learn what we can from them."

Cecilia asked, "shouldn't you release them on Eden?"

George shrugged, "we could, but I'm not sure that's smart. We can presume that they, normally, have organochemical flexibility in order to prey on whatever they can find during an ecological succession. But, normally, they wouldn't have any exposure to Earthly biochemistry or ability to digest Earthly compounds. If we release them, do we risk propagating the ability to consume Earthly compounds among Eden's creatures? It's not worth the risk…"

There was a long pause, then Elke nodded, thanking him for his report. She turned to Nikko, "have you been able to compile any estimates of the toll?"

Nikko paused, and leaned forward, fingers meshed on the table surface. He cleared his throat, looked about the room, and spoke solemnly. "We are still doing well checks, but we have, at last count, 137 colonists unaccounted for, in addition to our own casualties among the colonial staff."

There was a long pause.

Elke did not take her eyes off her coffee mug. "Please continue your well checks. Please, be thorough, do your best to find people that are unaccounted for. We'll hope for the best. But, as soon as humanly possible, I would like a list of names of the lost. We will need to hold memorial services."

She paused, then continued, "please, also, when you have a definitive list of the lost, include their skill sets. It will be a year before the next ship arrives; we'll need to understand whether we have any critical gaps that we'll need to address through cross-training." She paused again, then looked up and asked, "what is the status of the colonial rings?"

"There's very little structural damage," Nikko continued in monotone, "mostly from the vigilante explosion and fire. We've lost a great deal of livestock. Our preliminary estimate puts our losses at eighty percent of the goats, eighty-five percent of the chickens, and almost ninety percent of the pigs. It's possible we'll find a few that escaped their pens and are now roaming the ring; I'll update when we know for sure. About fifteen percent of our crops were lost in firefights with kraken. Most of the colonists' dogs and cats are gone." He sighed deeply, then continued, "our diets are going to be simpler going forward; we have to give livestock populations time to recover. Fortunately, the kraken did not develop a taste for our fish, and we can still produce fabricated proteins from our algae farms."

Elke nodded, then raised her hands, "any bright ideas as to how to determine whether there are any surviving kraken?"

After a pause, Dani spoke, hesitantly. "I'm…I am sorry for saying it this way, but, if there are any more, well, we'll know when someone is attacked. I don't know what else they'd find that's big enough to prey on, at this point, other than colonists."

"Well, there's also the motion sensors," Moira commented. "We've positioned them throughout the residential rings' plenum spaces. We'll know if any kraken are in those areas."

Dani asked, "What about motion sensors in the agricultural spaces?"

"Incomplete," Nikko jumped in. "We pulled the teams out when we engaged the mating kraken that attacked our security team, and then formed the line. There are some spaces that have not yet been instrumented."

"Let's take care of that, as soon as possible," Elke said. "Let's end the line advancement and finish deploying the motion sensors." She paused, then asked, "Moira, can you equip them with electromagnetic detectors like the ones we placed in the drones? We don't want them surprised up there."

Moira nodded firmly. "Give me an hour or two."

Elke nodded. "Dani, should we give the line an hour's break, then prep them for finishing the sensor deployment?"

Dani looked at Nikko, then replied. "Sounds good. I'll keep the drone flying, though, just in case."

▪ 08:31

The gravid kraken was nearly weightless here, requiring it to grip tightly as it made its way through unfamiliar space. The dark was complete; it was unable to see anything at all. It could sense the movement and electromagnetic signatures of very small creatures as they scurried away.

It felt compelled to keep moving, to get as far as it could from where it had been feeding. Eventually, it came to an area with unfamiliar air flows. It could not feel the electromagnetic signatures of prey, but the air carried scents that it knew, not too long ago, would have been compelling. Now, though, it just wanted to rest in a safe place.

The kraken explored the dark space as best it could, all of its senses straining to detect danger of any kind.

Nothing.

It found a place that seemed secluded. It was dark, and it was able to back itself into a tight place. Wedged between hard surfaces, the kraken kept its claws and beak exposed to protect itself as needed. Listening to the total silence, it stopped moving. It felt the rhythms of its own breathing. It could feel the pumping of its own blood. It could feel the movement of its young as they became increasingly active. It held fast to the structures around it, retracted its sensory cluster, and, for the first time in its life, shut down to an unconscious state.

▪ 09:14

The diploid young awoke, for the very first time. It was in a dark, warm, moist place. As it moved, it felt the movement of its siblings, in the confined space they shared.

It felt intense hunger, greater than it had ever felt before. It reached out with its forelimbs, stretching out its long, thin, digits. Each ended in a short, sharp talon. It grabbed the nearest moving thing. It opened its beak and stretched out its tongues, grasping its sibling, pulling it into its mouth, and began digesting it despite its struggles. It tasted wonderful, and it could feel its siblings' pain and fear. It was all very new, and very exciting.

As it consumed the smaller one, it began to realize that others were being eaten as well. It did not want to be eaten, so it finished ingesting its sibling as rapidly as it could. Then, it grasped another, and repeated the process, growing very quickly as it did so.

Soon, there were only five left. It did not feel confident that it could subdue any of the others. It stretched out its claws and opened its beak, and began tearing at the membrane holding them.

It tore and chewed its way through the membrane, then through the muscles and fibers beyond the membrane until, finally, it reached a point beyond which there was no further resistance. It pulled itself out, moved a short distance, and exhaled the fluid from its lungs. Then, for the very first time, it inhaled air. It began moving across the body from which it had come, sensing the others following it through the hole it had created.

It felt a tug as it moved, a tug from something connected to its belly. It was a flexible tube, extending back to the opening which it had come from. It bit the tube with its beak, severing it. It resumed moving across the body as the bit of tube still attached to its belly shriveled, then fell away. It paused to consume it.

The body beneath it was breathing, but not otherwise moving. Somehow it knew that the body would never move again. It moved away from where it had exited the body to avoid the others that would soon be coming out. Then, stopping when it felt safe, it began to tear at the immobile body with its beak and talons, using its tongues to pull the bits of flesh into its mouth. It was hungry, and the body was delicious.

▪ 10:37

The diploid young continued to feast on their parent's remains. It was not clear at what point the parent's life had ceased, nor was it important. The parent had served its purpose. They were young, strong, well fed, and ready to hunt.

As the parent was reduced to a skeleton, they became increasingly aware of each other. When the last morsels were torn from the parent's carcass, the largest spread its wings and talons and issued a long, low, guttural call. The second largest briefly considered challenging its womb mate, then decided against it and lowered its beak. The others followed suit.

The largest, the alpha, folding its wings, walked to each in descending order of size. It opened its beak, then gently closed its beak over each of its womb mate's sensory cluster. It was not enough force to cause damage, but just enough to establish dominance. When it came to the smallest, it hesitated, then decided. When the largest closed its beak over the smallest one's sensory cluster, it bit down hard. Then, grasping the smallest with its outstretched talons, the alpha tore off the sensory cluster and consumed it. It continued to methodically dismember and consume its womb mate, who obediently submitted without resistance. It had accepted its fate.

When the smallest one was completely consumed, the largest raised its sensory cluster, pulled it closer to its body, spread its wings and took flight.

It followed the air movement to one of the nearby ducts. Its wombmates followed obediently.

▪ 10:45

The sweep team of security and operations people were resting, sitting or sleeping on the grass of the agricultural ring. More of Nikko's people arrived bearing the portable electromagnetic field detectors Moira's team had fabricated. They sorted out who would carry the detectors and who would carry the motion sensors; when they were set, they made their way to the ring's edge and plenum access.

The senior staff in the conference room were also resting on cots and mats brought in by Nikko's people. They stirred and woke to their individual alarms. Elke went to the corner fabricator for a cup of coffee. It wasn't as good as the real brew she indulged in when breakfasting in her quarters. But she wasn't that particular at the moment.

She walked to the window and looked down at the planet's surface; it was still incomparably beautiful, despite the horrors of the last few days. The area beneath them was dark; she had to check her wrist computer to determine whether the day below had just ended or was about to begin. She turned to look about her conference room turned command center. She saw her team stirring. Some were still groggy, some padded to fabricators for their preferred stimulants. She saw the flamethrowers and scatter guns, delivered by Moira's engineering team, propped in the corner after they'd completed their check-outs. She saw the isolation chamber in another corner, still holding the dissected reproductive organ taken from one of the kraken.

Along the walls, she could see support personnel. Nikko's people folded up the sleeping cots, seeing to the needs of the senior staff. Moira's people fiddled with one of the display consoles that continued to misbehave throughout their crisis; one of them lay on her back, head in an access panel, reaching up into the space below the console. Some of George's people worked at another console, quietly debating and gesturing at a projection of a walking kraken. She thought she should go listen in on whatever they were debating, but not just yet.

She sipped her coffee, feeling inexplicably good. The worst, it seemed, was over. She shook her head, then stared out the window again. The toll was horrible: 137 missing colonists, significant casualties among her colonial government team, injured survivors in Sick Bay. The limited damage to the station would be repaired. The lost livestock and crops would have economic consequences for the farmers, but they would eventually bounce back. Perhaps she should allocate some of the colony's income stream from precious metals and organic chemicals to a relief fund. That made sense. Despite her frustrations with Jenna, it would probably be smart to name the relief fund after her. She'd need to work through who would take her place as colonial lead and liaison.

But the people could not be replaced. Whole families were gone; other families faced devastating losses that would haunt them for at least a generation. So many were gone.

She looked down, then back out the window. She couldn't deny that she felt, in addition to the sense of profound loss, a satisfaction, pride even. As awful as these past few days had been, they had prevailed. Her team had come together and performed in the face of a dire and unforeseen threat. She took another look out the window, then walked back to her place at the conference table. It was time to take charge again.

▪ 10:51

The alpha and its womb mates flew through the dark, following the movement of air and the tantalizing scents it carried. They came to an opening, lighted around it, and sampled the air for scents. They reached out to detect electromagnetic fields. They moved downward in the dark, following the scents, through a passageway too small to fly through. So they climbed downward, the alpha in the lead.

▪ ▪ ▪

Elke sat at her place at the table. The others followed suit. Dani spoke: "the security and support personnel are ready to enter the plenum above the agricultural space and install motion sensors. The portable electromagnetic sensors are on. Nothing has been detected."

Elke nodded. "Proceed then."

▪ ▪ ▪

The alpha reached the end of the passageway and searched with all of its senses. The space was dark, but not the complete darkness they'd come from.

321

There was some dim ambient light. The space was open, but there were structures impeding their ability to fly freely. They made their way through, grasping the structures as they passed.

They could sense the proximity of prey. There was a large cluster beneath them, probably enough prey to sustain them. They knew, somehow, that their lives would not be long, but they were too ravenously hungry to be particularly concerned about that. They knew they must feed, at all costs.

They were in a race, a race to grow. If they could grow large enough, they would experience a delightful, exquisite release. Release of what? They did not know, nor did they particularly care. An evolutionary instinct told them the release was more important than their individual lives. If they needed to sacrifice themselves for their siblings, they would do so. Nothing was as important as the release. The sacrifice of their lives would be nothing if it meant that one of them could experience the release.

They moved quickly downward through the dark place. They felt themselves get heavier; excitement built as they contemplated the prey whose electromagnetic signatures they sensed below.

They reached a barrier. The alpha extended its talons and probed the barrier. The panel shifted. Abruptly, it swiped the panel aside, and they could see the brightly lit space beneath them. The alpha went first, spreading its wings just enough to slow its descent as it leapt down to the substrate in the bright space. The others followed, flanking the alpha as it began walking towards the prey.

They stopped, suddenly, when a prey stepped into the space before them, looking in the other direction. It was almost as big as the alpha. It did not turn its sensory cluster; it did not see them. It began walking in the opposite direction. The alpha wanted to experiment with the prey, to test its ability to suppress and attenuate.

But it was so, so hungry.

Partially spreading its wings, it leapt forward, with talons outstretched, beak open, and tongues reaching for the prey. It was nearly silent as it did so; but something, the slight noise, a movement of air, something alerted the prey. It stopped, turned, saw the alpha leaping towards it and made a loud, shrill noise. It turned to flee just as the alpha reached it.

The alpha's mass and momentum took the prey off its feet and down to the floor. The alpha's talons grasped and crushed the upper torso of the prey and pierced its flesh. It folded its wings as its hind legs landed on the substrate and one of the prey's legs. Its beak was open, and the tongues stretched out urgently, grasping the head of the struggling prey as it continued to call out. The alpha's tongues pulled the prey into its beak; the alpha pushed its head into its digestive organ. The prey felt its head suddenly pressed into moist, firm muscle. Unable to breathe and beginning to lose consciousness, the prey survived just long enough to feel the digestive fluids begin to flow.

The alpha was nearly vertical, supported by its two hind legs and two tails, wings folded against its sides. Its taloned forelimbs held the now still body of the prey as the tongues alternately stretched and pulled the prey deeper into the alpha's gullet. It took mere minutes for it to consume the prey as the others watched. Their hunger was not abated, but they could share the alpha's experience through its electromagnetic pulses, all reveling in the flavors and struggles of the prey.

The alpha leaned forward and set down on all fours to resume walking. The others followed close behind.

▪ 13:03

Elke paused. She thought she heard a noise. She shook her head, and resumed watching the projection of the agricultural ring plenum as the team moved through it, installing motion sensors near the screened air ducts.

The sweep team neared one of the main access ducts. The lead tapped his wrist computer and asked, "Daniella, can you see this?"

Dani used her hands to enlarge the projection. She repeated this two more times until they could see, above the conference room table, a square portal 2 meters wide with a wrecked spore filter.

"What are you seeing?"

"The screen has been shredded and pulled aside. I'm thinking a kraken might have torn through it and went up into the life support core."

Nikko nodded, "yes, that's exactly what it looks like." He turned to Elke, "they're not finished deploying motion sensors, but I think they should follow the access path to the life support core and look for that kraken."

Elke glanced at Dani, who nodded. "Do it."

Nikko said, "proceed into the access pathway and search the life support core. Stay together. We'll turn on the lights."

They watched anxiously as the team moved up into the low gravity of the life support core. They exited the access pathway and began searching through the space, moving cautiously in near weightlessness, relying on handholds.

Elke asked, "are they okay moving through there?"

"My people are," Nikko responded. "They're used to going in there to support Moira's people when they're servicing life support systems. But I can't speak for Dani's people."

Dani shrugged and said, "my folks aren't strangers to low gravity, but they're not as proficient."

They continued to search. Then, there was an exclamation.

Everyone in the conference room slowly, and without thinking, stood as the team moved to the end of the life support core above the command rings. As they did so, and as the model shifted perspective, they all could see the skeleton of a very large kraken wedged between air handlers.

"Shit!" George grabbed the sides of his head in disbelief and frustration. "This is not good." He looked about the room wildly. "Okay...so, once again, this is speculation, but I can't think of an alternative explanation for why that kraken has been reduced to a skeleton, other than...its diploid young ate it, and escaped. It'd be one thing to find the carcass intact, but finding it skeletonized means something fed on it. I think this means one or more of the diploid young have emerged and consumed their parent's carcass. That would provide them the biomass to grow substantially. Now, they're somewhere in the colony, and it's only a matter of time before they start releasing spores."

Another exclamation came over the display, and a voice said, "there are droppings over here by the access pathway to the command ring. I'm guessing it, or they, went through here! Whatever it is, it's headed YOUR WAY!"

Elke turned to Dani and Nikko and shouted, "distribute those weapons NOW!"

She looked about the room, pointed, and said, "If you have not had any weapons training, get to that corner and STAY OUT OF THE WAY!" The junior officers and untrained crew rushed over to the corner to take cover. Elke, George, Dani, and the other senior staff quickly armed themselves.

Elke turned to the display. "There aren't many people in the command rings, other than the people in this room. I can only assume these diploid young have similar sensory capabilities to their parents. That means it's only a matter of time until they sense us and head this way. I want you to follow them down that shaft. If they come after us, you can come up behind them. But be careful!"

George put his arms through the shoulder straps of a flamethrower and yelled, "they need to be very careful! The kraken hunted individually; we should not assume these things will behave similarly. They may be in the plenum area above the command ring!"

Elke nodded and asked the display, "you heard that? Good. Stick together and pay attention to your motion sensors!"

She strode to the corner and put her arms through a flamethrower back as well, then picked up a scatter gun that she attached to the pack with a clever clip Moira's team developed.

She looked above her. "Is that ceiling secure?"

Nikko responded, "no, actually. The panels are on a frame; they're not attached."

"That's just *great*," David muttered.

Elke surveyed the situation. "Then we need to stay in this room so we can protect the unarmed." She glanced about the room, "I don't see any part of the room that's more defendable, so we'll keep you all in that corner. The rest of us should spread out. If they come through the ceiling, burn them."

"Expect the flamethrowers' range to be up to two meters," Dani cautioned. "We don't want to burn each other."

George pondered aloud, "we could try the scatterguns. They may be more susceptible than the kraken generation."

"The scatterguns don't have a meaningful range limit," Dani countered as she donned her pack , "even though they do fire a conical spray of rubber bullets. I think, under these circumstances, the flamethrowers may be safer."

"Agreed, flamethrowers only," Elke commanded.

"Wait!" Everyone spun towards Nikko; the look in his eyes gave Elke chills. Nikko looked at her:

"What about Sick Bay?"

▪ 13:51

John sat in his office chair, reviewing and updating records and contemplating treatment plans. It seemed incredible that it had only been three days since the discovery of Esteban's remains. The patients that had their parasites surgically removed were recovering as well as could be expected. Nanobots were rebuilding limbs and repairing neural tissues. He was the most concerned about the severe burn victims. A century before they'd unquestionably have died. Today, he can keep them alive, but, for the worst of them, he wasn't sure that was a kindness.

He leaned back in his chair and pinched the bridge of his nose. He needed to rest, but there was no time. He hoped, at least, that no more patients would be coming his way.

AAAAAAAAAAAAH!

A scream and a crash brought him out of his reverie and to his feet. He reached the door to his Sick Bay and stared.

Nightmares were in his Sick Bay! Nightmares were attacking his staff and patients!

The largest was easily three meters long and 200kg. It stood vertically on stout hind legs and a pair of short, thick tails. Wings folded against its flanks, and it rapidly pulsed waves of bright colors. Forelimbs with six digits, ending in long, sharp talons, held Natalia suspended in the air. Blood dripped where the points of the talons pierced her skin.

She was nearly horizontal, facing the beak of the thing; she screamed and struggled to no avail. The thing had a sensory cluster like the kraken; its six eyestalks stared into her face. Its beak opened, and nine long, muscular tongues extended and, in moments, stripped everything from her body. Then it stretched out to grasp her and pulled her headfirst into the beak. He heard her awful screams muffle as her head was pressed into its gullet. The tongues continued stretching and grasping and pulling, as her legs kicked wildly. Finally it released its claws, raised the angle of its beak, and finished swallowing her.

John swore he could still hear her muffled cries for a few moments after she disappeared. Another commotion caught his attention. Its smaller counterparts were attacking his other patients.

One used its talons to hold down a screaming, struggling, legless patient. Its hind limbs perched on the floor; its tongues stripped the patient. It turned, used one forelimb to lift the patient's upper body, then extended its tongues and began pulling him headfirst into its beak. Another used its talons to shred apart one of his burn victims, pulling bits and pieces into its beak with its tongues. Eventually, it consumed her the same way.

The last held one of his nurses down on the floor. She was face down, shrieking, as it stripped her of her uniform and pulled her head into its beak.

One of his nurses saw him and ran towards his door, but was caught in the outstretched talons of the largest monster. It lifted her high, stripped and swallowed headfirst. The poor nurse barely had time to scream.

In a matter of moments, all of his patients and staff were consumed. The burn victims were gone. Their recovery beds lay empty, except for shreds of equipment that had been used to heal their bodies. The kraken attack victims were gone. Jian was gone. Corinna was gone. The largest turned and saw an orderly, sobbing and pleading, cowering in a corner. It grabbed her and subjected her to Natalia's fate.

It all happened so fast.

There was no one alive to be saved. John collapsed against the doorframe. A large lump welled up in his throat. Wracked with guilt and suffering from shock, he stepped back and secured the door, rendering it opaque. He had no idea whether these things would stop by his door; he prayed they wouldn't realize that they could simply climb over the wall through the plenum.

He tapped his wrist computer frantically. "Elke! Elke, they're in Sick Bay!" Her voice responded, "John? What's in Sick Bay?!"

He sobbed, "I don't know! They're like the kraken, but they're *different*. They're like dragons. They're like winged DRAGONS!"

"We're coming John! Are you okay?"

Staring at the door, he whispered, "yes, but I don't know for how long."

She sounded out of breath. "Are your staff and patients safe?"

"I don't know how many got away. I saw…" He choked, "I saw them kill my patients. Oh, God, Elke, *I watched them eat my patients!*" John dropped his head between his knees and sobbed. He sobbed, as quietly as he could. He couldn't stop crying; instead, he prayed the monsters beyond the door didn't hear him.

■ ■ ■

When John called, Elke turned to George and instructed him, "I want you to stay here with your team to defend the unarmed!" She resumed listening to John. Her face became incredibly pale, whether with anger or fear George could not tell. Her hands squeezed the flamethrower wand so tightly that he thought she might break it.

After John went quiet, George grabbed Elke's arm. "Elke, they must be the diploid offspring. We've got to destroy them before they release spores!"

Elke didn't respond, but turned to Dani and yelled, "tell your team that's coming from the plenum to go to Sick Bay, *now*!

The rest of you, COME WITH ME!"

▪ 14:09

John stood in the corner of his office, unsure what to do. He scanned the room. There were no alternative exits, save the plenum, and no objects that had any potential as weapons. He looked at the plenum and began to weigh his options. If the nightmares in his Sick Bay were aware of the plenum, he had no hope anyway. He looked up at the ceiling; his one escape route was out of reach. He looked about the room again. Then, as quietly as he could, he pushed his desk into the corner, and began stacking storage boxes in a pyramid on his desk.

It was a daunting task. The ceilings were five meters high, but the alternative was grim. He began to climb the boxes, grimacing with every sound and wobble. Finally, he was able to reach up and slide a ceiling panel aside.

He curled his fingers over the frame that held the ceiling panels, said a small prayer, and did his best to coordinate a small jump with a pull. Thankfully, his stack of boxes did not shift. He managed to get his chest over the frame; his legs now dangled uselessly above his makeshift ladder. He extended his right arm and searched in the dark for a handhold he could use to pull his lower body into the space.

He found one. He took a deep breath and pulled himself into the plenum. He stood, let his eyes adjust to the darkness, and quietly replaced the ceiling panel.

▪ 14:16

The alpha looked about the enclosure and was pleased to see its siblings feeding and growing. It could see the swelling in their abdomens, and it could feel the swelling in its own belly. The time was near. It was already too heavy to fly; soon it would be unable to move.

It scanned the room visually and through its electromagnetic organ. The only remaining prey were being consumed. It felt more signatures in the distance, but the direction was unclear.

It had seen another prey briefly enter the space and walked to where it had disappeared. There was an irregularity in the enclosure boundary; an electromagnetic anomaly without anything that corresponded visually.

It probed at the anomaly with one of its forelegs, then began methodically searching the surrounding area. It leaned back on its haunches and probed at the barrier above it, noticing that the barrier was not firm. It knocked aside that barrier and recognized the dark space above that they had traversed earlier.

It reached out with its senses and did not detect anything nearby. It considered climbing into the dark space, concerned that the electromagnetic anomaly was interfering with its perceptions, then decided to continue hunting in the lighted place instead.

It could feel a number of electromagnetic signatures heading their way from different directions. The alpha and its womb mates had already grown considerably. They needed more prey to prepare for what was coming, but nothing, nothing was as important as survival. It was anxious about the number of signatures heading its way.

The alpha decided to send its smaller womb mates first.

▪ 14:27

Elke led her contingent through the corridors between her conference room and Sick Bay. She coordinated with the security lead from the life support core, ordering them to approach Sick Bay from the opposite direction. She understood the risk that the predators could return to the plenum anytime, but there wasn't much she could do about that.

As they approached, a figure stepped out into the corridor.

Elke stopped and exclaimed, "Jian! Are you okay?"

Jian, sobbing, shook her head violently. "They're in there! They're killing everyone!"

Elke grabbed her by the shoulders and pulled her out of the main corridor. "We know. Did anyone else get away?"

Jian shook her head again, "I don't know…I don't think so…I don't know." She broke down sobbing again.

Elke, knowing the security team would be approaching Sick Bay, pushed Jian towards one of the operations people. "Anju, take Jian back to the conference room and stay there to help guard the unarmed."

She left Jian in the guard's hands and resumed walking.

▪ ▪ ▪

As they approached Sick Bay, it stepped out into the corridor. She immediately understood why John called it a "dragon." It was vaguely reminiscent of a kraken, yet clearly different. It was longer and leaner, with a beak and a sensory cluster; but instead of claws retracted under its body, it walked on forelimbs ending in long, curved, cruel talons. Its middle limbs were wings, folded against its flanks, and its hind limbs were stout and strong.

As it turned towards them, its flanks flashed bright colors. It took a few tentative steps their way, then was followed by another that turned in the opposite direction to confront the security team.

Elke began to feel tired, a little muddy. She realized it was subjecting her to some dampening electromagnetic impulses. She shook her head and tried to concentrate.

Suddenly, it began running towards them. As it gained speed, its wings unfolded slightly, helping it run on its hind legs. Its forelimbs and claws stretched out towards her team.

"Form a line!" Elke yelled.

As she stepped forward, Dani, Moira, and Nikko flanked her. When the creature was within range, they activated their weapons and drenched it with fire. It continued charging as they backed away, the others stepping back as well. They could smell the fuel of the flamethrowers and the burning meat of the predator as it cooked and burned. It reared up on its hind legs, its wings spread and its claws reached towards them. It slowed, continuing to step forward, drenched in flame. Finally, it faltered, and fell.

Elke had to yell over the buzz of the flamethrowers. "STOP, George will want to examine it."

Her wrist computer vibrated. She tapped it and heard the voice of the security team lead.

"We just killed it. Whatever it is, it's dead."

She panted from adrenaline. "We've killed one, too." She quickly turned and surveyed her team. Dani, Moira, and Nikko each flashed a thumbs up sign, showing they were alright.

After a moment, she pointed beyond the smoking carcasses. "Let's get to Sick Bay!"

They stepped carefully around the smoldering remains and continued cautiously down the hall. They neared Sick Bay at the same time as the security team. Elke turned to the security team lead, "any idea whether there are any more?"

The security lead responded, "we came across a couple of nurses that managed to get away. One of them said there were four. I believe we've accounted for two."

Elke nodded, "they're either in there waiting for us or they escaped into the plenum."

The security lead nodded and looked over at the doorway. "Your lead, Governor."

Elke nodded, then turned to Dani, Moira, and Nikko: "I'll step through first; I want you to follow me and immediately form a line inside."

She stepped through the doorway and heard her team follow. The creatures were gone; ceiling panels were knocked away.

She relaxed slightly and turned back towards the security lead. "Bring the nurses in; they can help us assess what's happened."

The security lead entered the room, his hand at the elbow of one of the nurses. The nurse, shaken, let out a squeak. "They're all *gone!*"

Elke approached and stood directly in front of her. "Help me understand what happened here."

The nurse nodded and swallowed hard. He pointed to the left wall. "The burn victims were along that wall, with the kraken attack survivors. That's where we have the regenerating equipment; they were being treated with nanobots that were stimulating tissue regeneration. They…" The nurse trailed off, sobbed, then continued, "they couldn't run; they just laid there, screaming, as the things tore into them."

The nurse pointed to the right wall, "that's where the surgical recoveries were. They were in beds, but they were mobile; some still had IVs. The big one went after them first." He sobbed again, then added, "I was near the door. When they rushed in and attacked the patients, I just ran. I abandoned them…" He broke down.

Elke put her left hand on his shoulder and tucked a right knuckle under his chin, gently lifting his head until he was looking up into her face. Softly, she said, "you couldn't have helped them. If you'd stayed, you'd be dead now, too. You did the right thing; you're helping all of us understand what happened." She released his chin, then continued, "it's critical that you collect yourself, and tell us everything you can remember before these things attack anyone else." The nurse nodded, then, with a big sigh, continued:

"We had no warning. They rushed in, through the door, from the corridor. The second one was much bigger than the other three. They let the big one attack the surgery patients while they went after the poor bastards trapped in their beds. I saw it take Anna, one of the surgeons. It…It just grabbed her with its…" He paused, his mouth gaping as he sought the right words; he held out his own hands and stared at them. "It's *claws*. It grabbed her, and picked her up, and she was screaming, and it opened its mouth, and these, these, *things*, these…*tentacles* came out of its mouth and started tearing her clothes off and then wrapped around her and pulled her into its mouth…" His voice trailed off. "She kept struggling as it swallowed her, gulping her down." He shook his head and sobbed. "And that's when I ran."

Elke patted him on the shoulder; keeping her eyes on him, she issued a command to the security director next to him: "search the adjoining rooms, and see what you can find. We should search for other survivors in the corridors, but I don't want any small groups out there. Remember that they have that dampening capability. Make sure there's at least ten of you at a time, and that you're armed."

The security detail fanned out and searched the facility. After a few moments, the security lead came back. "I don't know how many people were here, but judging from what we're seeing, I'm guessing these…things killed some twenty people. And you should look in here." He pointed towards an office door, then continued, "we had to override a lock."

Elke walked through the doorway into John's office, and saw the pyramid of boxes in the corner of the room. She turned to the security lead and asked, "do you think he survived?"

"Can't tell. We went up and can see that he made it into the plenum, but we have no idea where he went."

Elke tapped her wrist computer and spoke softly. "John? Can you hear me?"

After a tense pause, his voice came through, "I'm here."

She smiled, "wonderful. Where's 'here'?"

"Ah, I'm rather high in the plenum, not far from Sick Bay. Are you in Sick Bay?"

"Yes, we are."

"On my way."

Elke turned and began to walk the room, taking stock of the carnage. There were bloody shreds of cloth, bloodstains on the beds, bits and pieces of medical equipment. She shook her head. She heard a noise above, and saw John's legs dangling down from the plenum. The security lead climbed the pyramid of boxes to help guide his feet to secure footholds.

When he reached the floor, John turned and asked, "how bad?"

Elke responded, "I was going to ask you that. We found Jian and a couple of nurses that managed to escape. I don't know whether any others got away; we'll need to do a search. What can you tell me?"

The others circled them as John spoke. "I heard a crash and a scream. I came to my door and could see the things attacking patients and staff. It was awful. In a matter of minutes, they'd killed and consumed everyone. I didn't see anyone escape from where I was. There was nothing I could do." He hung his head, then looked up, "I locked and obscured my portal, and did my best to escape. I figured, if the kraken were pretty smart, these things probably were too, and it wouldn't take them long to figure out they could get into my office through the plenum. So, I climbed up and did my best to hide."

Elke nodded, her lips pulled tight. "We killed two," she informed John, "how many did you see?"

John thought for a moment; he visibly shivered from the memories. "Four. One really big one, about three meters long, and three smaller ones."

"We must have killed two of the smaller ones," Dani pondered aloud. "There are ceiling panels knocked aside; we think the remaining two escaped through the plenum. Did you see them?"

John shook his head. "No. My hiding place wasn't directly above Sick Bay, so, no, I didn't see them."

Elke led John and the others out of the office; she readjusted her flame-thrower pack as she spoke, "This isn't over. They'll either resume hunting or release spores. Worst still, they might do both. We need to get back to the others."

John looked about, "I'll come with you. There's not much I can do here. My nurses and I will bring some supplies."

Elke nodded and started walking. Dani, Moira, and Nikko accompanied her and the rest followed suit. She looked up and down the corridors, "any way we can determine where they are?"

"Not here in the command rings," Nikko said, "or the life support core. The residential and agricultural rings are now instrumented with motion sensors. If they go there, we'll know."

"Good. Now, how do we get ahead of these things? I'm tired of reacting, of letting them make the first move. It's time to take the fight to them."

"I don't know," Dani answered cautiously. "We could wait in a fortified position and let them come to us; or we could organize hunting parties and go after them. I don't see other options, do you?"

▪ 15:06

The alpha and its womb mate waited motionless as their siblings attacked the approaching prey. They were ready to follow and enjoy the feast, or they could retreat into the dark space above, depending on the outcome.

When they sensed their womb mates had died, they chose retreat. The alpha reared up and pushed aside some ceiling panels; it then reached upward and grasped structures that allowed it to pull itself into the dark space.

The swelling in its abdomen was growing. Soon, it would need to find a safe place for the release. Its womb mate's abdomen was growing as well, but it needed more nourishment. Once they'd climbed far enough to feel secure, the alpha paused to weigh their options. It considered whether to simply consume its womb mate in order to ensure release, or whether to send it in search of food in order to ensure one of them would survive long enough.

After a moment, it made a decision.

The alpha resumed climbing through the dark space to where they'd been born. Its womb mate descended, monitoring the movement of prey below. There were now two clusters of prey: one stationary and the other moving towards the first. The alpha continued its descent into the lighted area between the two groups. It found a shielded place and backed into it. It reached out with its tails until it found a way to anchor itself. The swelling in its abdomen grew, as did its hunger.

It sensed that one of the prey from the stationary group was moving towards it. It matched color and texture to the materials around it, and began gently bombarding the prey with electromagnetic signals. It experimented, noting what did and didn't work.

▪ 15:23

Leonard was uncomfortable. They'd been in the conference room for some time now, and nothing was happening. He'd overheard the xenobiologists discussing the two things that had been killed in the corridor. Apparently, there were two more unaccounted for.

He really needed to use the restroom. He stood and walked to the lead xenobiologist. As he approached, he said, "George, I really need to get to the restroom. Sounds like the action is over, and it's just down the hall." He smiled awkwardly, then added, "I promise I'll be right back."

George hesitated, and looked around the room; he realized Leonard wouldn't be the only one. He grunted with unease, but eventually relented. "Alright." He raised his voice, "people, we're going to do restroom breaks. Two at a time, please."

As the others started talking amongst themselves to sort out an order of departure, Leonard approached George again. "George, I can't wait. It's just down the hall. I'll be right back."

George nodded.

Leonard strode purposefully towards the door as George turned to work with his peers. As he approached the door, he tried to shake off his sudden grogginess. He kept walking. He exited the room and slowed as he walked. He suddenly stopped, and, as he did so, he turned to the left and realized that the panel was moving. In fact, it was reaching out for him.

Nine long tongues stretched out for him, their tips sliding wetly around the back of his neck, his shoulders, his waist, and his legs. As they curled, he observed their heavy musculature. They were exquisitely smooth as they gently stretched across his skin, almost caressing him. Then they suddenly became rough and sharp, digging into him as they tensed and began to pull. It was very painful, but somehow he didn't mind. Something about it was disturbing, but he also felt wanted, almost welcomed. He felt desired; he was at peace. In fact, Leonard felt *grateful*. He smiled even as he shook his head. It was the strangest experience.

His vision was blurry, but he could see, behind the tongues, a dark, wet opening. The strong tongues pulled his head down and forward; He began to lose his balance just as two limbs reached out to hold him. He could feel the claws piercing his chest and abdomen; but once again, as painful as it was, it was also, oddly, pleasurable. He felt the movement of the tongues change. Leonard was suddenly aware that he was naked, but he felt no shame.

He was now horizontal, facing downward, his head pointed towards the opening. He felt the top of his head press into wet muscle. The rhythmic contractions of the muscle gained traction as he felt his face pass into it, then his shoulders. He could not breathe, but he wasn't concerned. He wasn't afraid. He continued smiling even as he slowly began to suffocate.

▪ 15:54

Elke turned a corner towards the conference room and felt very off. She looked towards the conference room and realized her vision was blurry.

She saw movement, but she could not quite make out what she was seeing. She stopped and shook her head as the others continued.

The security lead was first. He walked towards the door, then slowly stopped. As he stopped, she saw the panel reach out for him. Elke swayed on her feet; she was terribly dizzy. As she stared, nine long tongues emanating from a splayed beak grabbed hold of the security lead, as two clawed forelimbs reached out from the opening in the corridor wall.

She fought a battle of will. One voice whispered within her, *everything is fine, everything will be fine.* In fact, she felt good. Soon it would be her turn. She looked forward to her turn. It would make her...*happy*...

NO! Another voice screamed within her, telling her to resist, to fight off the sleep. Her people were in danger. Her *friends* were dying, and she stood there, doing nothing. This was not good, this wasn't right. Her body tingled and her head throbbed. The pain reoriented her. The pain meant danger. Her fingers twitched around the flamethrower wand. She needed to *act*, and act *now*. She had to save her people!

With agonizing slowness, she snarled as she unslung her flamethrower. She stepped slowly forward, with great deliberation, taking one desperate step at a time. As her security lead's smiling face disappeared into the beak, she took another step, then another, until she was standing before the panel.

She was dimly aware that she was the only human moving. The security lead's feet flexed as his legs disappeared into the wet void, and the tongues stretched out for her. Her head throbbed so painfully her legs shook from the waves of nausea and agony. Shakily, she brought the wand of the flamethrower up. Tears welled up in her eyes; she was tired of fighting. She wanted to drop her weapon, to walk forward into its embrace. She longed for its touch, for the tongues and claws to reach out and take her home. There, inside its beak, she would finally be safe and warm and secure.

Don't...

Its eyed fingers danced in the air as the dragon reached out for her.

Give...

The tip of the tongues tasted her cheek, as a single tear streamed down her face.

IN!

Elke held up the wand, wrapped her finger about the trigger, and pulled. She pulled the trigger and wept. She wept because it hurt, because she longed to be cared for, to be relieved of her worries, all her cares, all her pains and anguish. And she wept because she couldn't stop, even as a voice shouted inside her. Hot, hissing flames shot forward against the panel, and the panel came alive. Her mind cleared, and she screamed. She screamed in fear, in anger, demanding vengeance. Behind her, she sensed the others come back to life; Dani, Moira, and Nikko joined her. Together, they set the awful thing aflame, until finally, it collapsed into a burning pile.

Elke suddenly felt a tugging at her elbow. It was George. "Elke, Elke, please, I have to examine the carcass. Please leave me something."

First, it had been hard to pull the trigger. Now she couldn't imagine releasing it. As she released the trigger, reluctantly, she ordered the others to let up as well.

She stared at the smoking remains. "How…how did it do that to me?"

George looked at her quizzically, "what do you mean?"

She paused, then pointed at the remains. Her hand shook violently. "I couldn't move. And, when I managed to get in front of it, all I wanted to do was let it take me." She shook her head and turned away. She wiped away the tears furiously. "Pulling the trigger was the hardest thing I've ever done."

George reached out and patted her shoulder. "We all felt it. The electromagnetic suppression seems to be even stronger in this diploid generation than the kraken."

John, having walked up while they were talking, leaned slightly forward and asked, "how do you feel now?"

She looked at John. "Shaky, and I have a bit of a headache."

He scanned her quickly. "We need to get you checked into Sick Bay. We need to get *everyone* checked in Sick Bay. There might be some brain damage."

She pulled away from George and stared down at John. "Right now? No. No, right now, my brain is fine. And there's one more of these things somewhere in our colony. We need to find it before we do anything else."

She nodded over her shoulder to the diploid carcass. She couldn't bear to look at it again; she still had too many confusing emotions. "I need to know how close that thing was to releasing spores."

George nodded, beckoned to some of his team, and got to work.

▪ 16:27

The alpha moved through the life support core, searching for a suitable spot. Its abdomen was swollen profusely; it couldn't keep moving much longer. It wanted more prey; it wanted to keep growing. It wanted to make the release all that it could be, but it knew it was big enough for the task. Further risk was not acceptable.

It searched the space from one end to the other. Finally, it found a suitable spot: a sheltered alcove not far from the space its parent had chosen for its release.

It backed into the empty space and extended its tails until it established a firm hold. Despite the absolute darkness, it managed to match the colors and textures it could barely detect. It extended all of its senses. Soon, it would be unable to move; but should anything come by, it was ready to hide and, perhaps, to feed.

▪ 16:56

George conferred with his team, then came back to Elke and John, who was still trying to convince her to go to Sick Bay. "As always," George said, "I'm doing my best with limited information. The organ we think is the sporangium is significantly larger in this individual than either of the two you burned in the corridor by Sick Bay. I'm taking some tissue samples to view under magnification, but I think what I'm seeing are spores. If the other one is similarly developed, we don't have much time."

"How many spores?" Elke asked through gritted teeth. She still had a nasty headache from her fight with the diploid.

George looked back over his shoulder. "The sporangium had grown tremendously, compared to the rest of the body. I think it was tucked into this space because it was having difficulty moving. But as to a spore count? He looked into her eyes, "I'm guessing, but I'd say it had millions of spores ready to go."

"FUCK!" Elke spat from anger and pain. She turned, unsteadily, towards Nikko and Moira. "Is there ANY way to determine where the last one went?"

They looked at each other, then Moira turned to Elke. "No, I'm sorry. We can take an educated guess, but that's about it."

Elke wiped away the sweat from her forehead. "I'll take the guess."

Moira, Dani, and Nikko formed a semicircle and lowered their heads. Nikko began, "We have motion sensors in the residential and agricultural rings, so it didn't go there. Could be in the plenum above us, but I doubt it."

"It could be in the commercial or industrial space," Dani continued, "or it could be in the life support core. I'd start in the life support core, then spread out from there."

"Why the life support core?" Elke asked worriedly.

"First, it's a central location," Moira answered, "so it makes sense to start there and extend searches from there. Second, it's quiet and dark because we shut the lights back off after we found the kraken skeleton. The darkness might appeal to our guest. Third, that's where it was born, what, a few hours ago?" She shrugged, "Fourth, if it's anywhere else and starts releasing spores, we can lock down that area, then open airlocks to purge. Awful if such purging impacts people or livestock, but at least it's a limited purge. If it's in the life support core and we don't kill it before it releases spores? We're all dead. So, yeah, I'd start there."

Elke barked a short laugh, then shook her head. "Hard to argue with your logic. Let's start in the life support core." She looked about and raised her voice. "EVERYONE! We need to find and destroy the last of these things. We're going to the life support core." She turned to George, "you and the other xenobiologists with weapons stay here to defend the rest."

Time to lead the charge.

She turned to Moira. "You should stay behind. If we fail, if we need to isolate and purge ring segments, you'll need to be in place to do so."

Moira shook her head with a tired smirk. "Not a chance."

Elke turned to argue, then Nikko touched her shoulder. "Between Moira's team and mine, we have enough people that aren't weapons trained that can manage that. We'll leave them in charge. You need more armed people on the hunt."

Elke blinked rapidly, then nodded vigorously. "Do it."

As Nikko and Moira walked to their teams to relay instructions, Elke turned to the rest:

"If you have a weapon, you're with me. We start in five minutes. Do what you have to do, and be ready!"

• 17:32

Nikko led them to the nearest access panel and pressed the controls. Ladder rungs popped out.

John laughed sadly, "that beats stacking furniture!"

Elke laughed under her breath. She turned to the team, "we're going up into the plenum and then on to the life support core. If you begin to feel 'off' in any way, if you feel sleepy, fuzzy, euphoric, anything at all, *say something*. I'm not losing anyone else to these things. Understood?" There were nods all around.

She turned to Nikko and asked, "you say the lights are off in the life support core? Can you restore them, to give us an advantage?"

"Yes," he responded. "I spoke with Dani. My people will very gradually increase the background light level while we're climbing, so it won't be completely dark when we get there. Hopefully, we won't spook our guest. Once we get through the access portal, they'll increase the lighting more, but not enough to blind us."

Elke nodded and reached for the ladder. Dani grabbed her upper arm.

"Let my team and I take the lead."

Elke paused, then nodded. Dani pointed to two of her security people, then pointed up. One aimed his flamethrower upward at the opening. "Anna, Dani, if you two see anything, drop to the floor and I'll cover you!" Both women nodded, then climbed up and took position.

The senior staff and their teams followed the security leads into the plenum. After their eyes adjusted to the reduced lighting, they followed the leads upwards as they began to climb.

When they reached the access portal leading to the life support core, Dani turned and spoke in a soft voice. "It'll be dark in the portal. Just climb the ladder. You'll feel a gradual gravity reduction as we go. It'll be disorienting, but you should be able to recover easily if you lose your grip. When we get to the life support core, my team will enter and secure the immediate area, then we'll initiate the search. It'll be a low-G environment, which some of you will find uncomfortable. We'll get through it together." She paused, looking at each of them. "*Please* be careful with your weapons. When you fire, you'd best brace yourself, or you'll jet away from your target and potentially injure your friends." She took a deep breath, then turned to two of her team, "Prasad, Anna, you have point."

They climbed carefully and quietly. Eventually, they could see a faint glow above them, coming from the dimly lit life support core. Dani's security leads stepped awkwardly through, followed by others with less low-G experience. Their movements were exaggerated, resulting in inadvertent leaps and loss of balance.

Elke shushed them. "Take careful, small steps. You'll get used to it; let your body learn."

The lighting levels increased. Soon, they could see their surroundings. There were no sounds.

Moira and Nikko knew the space best, so they led the way, following Prasad and Anna, flanked by Dani and the rest of her team. Elke and her team brought up the rear. They worked their way through the core methodically, starting with the end facing the planet above the command rings and essentially followed a long spiral through the dark space. Their initial tension began to lapse as time passed without incident.

They neared the far end of the life support core, above the agricultural rings. As they did so, Elke became disoriented. She shook her head. She looked at the others and could see that they, too, were experiencing something.

Dani grasped her head briefly, then looked back and caught Elke's eye. She nodded, acknowledging that they were, evidently, finally nearing the last remaining predator. Dani put a hand on Nikko's arm and, gesturing at her leads, stepped to the front. The three walked slowly forward, flamethrowers ready, and the rest followed.

Movement.

Dani and her leads froze. She stared into an alcove between large air handlers. She stepped forward, then stepped again. She flexed her trigger hand briefly, feeling the blood flow through her veins. Her finger tapped the flamethrower's trigger, and a tiny wisp of flame escaped.

Suddenly she was airborne! Two large claws grasped her about the torso as she screamed. She rose and plummeted abruptly as her captor, its body massive and swollen, awkwardly climbed above the air handlers. Her right hand wiggled the flamethrower wand as her right arm was pinned against her by its claws; her left arm flailed about wildly, pounding at the claws that held her. Her swears and screams echoed in the empty space.

The alpha reached the top of the air handler, and two of its eyes scanned the survivors as the remaining four locked onto Daniella. It opened its beak and the tongues shot out from the wet opening. They stripped her of everything, tearing at her flesh as it did so. As it attacked her, it released one claw's grip, then the other. This inadvertently freed her right arm, allowing her to swing the flamethrower wand into position.

"FUCK YOU!" she screamed, and opened fire.

It reared back, retracting its tongues, and tightened its grip on her torso. An awful, crunching sound rippled through the plenum. Elke realized it was Dani's ribs being crushed.

The shock of her flamethrower interrupted its electromagnetic suppression, and her security leads aimed their flamethrowers and fired. The entire side of the creature lit up with flames as they waved the wands side to side.

It backed away, gently falling in the low gravity. It tumbled, almost comically, from the air handler to the floor, still holding Daniella's body in one claw. Dani's team continued to advance as Elke, Moira, and the others joined them.

Suddenly, it threw Daniella's body at them, and rushed forward, claws outstretched. It grabbed and lifted Prasad with one forelimb; the talons pierced his chest as he continued firing. It reached out with its other taloned claw as the rest stepped back, but kept hitting it with unrelenting fire.

Finally, it slowed. Then it stopped, dropped Prasad, and fell to the floor. As it lay dying, a dark cloud wafted from a vent in its back. Elke screamed, "SPORES! Burn them!"

They shifted their attention from its body to the black cloud and set it afire. A few of Nikko's team focused their flames on the vent, burning the spores as they were released. The rest of Dani's team burned the hovering black cloud. Eventually, the thing's body collapsed as the last of the spores released.

Elke's voice shook. "Keep it up! Don't stop, keep burning it! I'm checking on Prasad and Dani!"

She walked to Prasad. John was administering first aid while talking with his team through his wrist computer. Without looking up, he said, "my people are coming. If they can get here fast enough, we'll be able to stabilize him and get him to Sick Bay. He should be alright."

Elke nodded blankly and muttered a breathy "uh huh." She looked past him to where Dani lay on the floor. She started to ask. John briefly locked eyes with her, before resuming his work on Prasad. He simply shook his head. "No."

Elke let out a noise, something between a breath and a scream. Her shoulders slumped, and the flamethrower wand dropped onto the plenum floor. She walked to Dani, and knelt beside her. The flamethrower wand made a light scraping noise behind her. Dani's eyes slowly locked with Elke's. She wasn't breathing. Her chest was completely crushed. She couldn't speak, she couldn't do anything but briefly look at Elke. Elke reached out and took her hand, holding it in hopes of offering some small comfort as her life ebbed.

Elke stayed with her until she was gone. Then, after a moment, she stood, and turned back to the burning creature. Her people were standing around it, their flamethrowers holstered, staring at its smoking remains.

As she approached, she asked, "have you seen any more spores coming out of it?

Nikko shook his head, "No. No, I think it's done."

Elke looked across the charred carcass. "Good." She looked at Nikko, "can you get the lights turned on here?"

He nodded, tapped his wrist computer, and the lights gradually brightened. In the better light, Nikko, Moira, and the others suddenly became aware of Dani. They all stood in silence, staring at her corpse. Some of them shed tears; others swore under their breath and kicked at the burning monster. They all were too tired to truly mourn her.

Elke looked just over top of Nikko and Moira. "Can you determine whether any spores got away?" Her voice was uncomfortably monotone.

Moira tilted her head. "I'm sure we got most of them, but I'm also sure we missed some. Unfortunately, given where we are, they'll be distributed throughout the colony. There's not a thing we can do about that, not without shutting off life support."

As she was talking, John approached them. A surviving nurse and a doctor kept working on Prasad. Moira continued, "I suspect, as the remaining spores make their way through the colony, some will land on vulnerable polymers, and we'll have some pinhole leaks. Those are easy to find and fix, and, as long as it's just a few, it won't be a problem. I'm more concerned about having to go through the colony and carefully clean every surface to make sure no spores come into contact with any of our people. We can't risk repeating this whole nightmare."

John wiped sweat from his brow. "Agreed. We should clean every surface, but, even then, we have to consider the possibility a spore will settle on someone's exposed skin. We'll need to regularly scan everyone, starting with ourselves, to make sure none of us have come into contact with a spore."

Elke nodded. The fire was gone from her eyes; her face looked sunken in, almost as if she were ill. She spoke slowly, tiredly. "Yes…let's err on the side of caution…yes that would be good." She looked up, meeting their eyes, then looked past them into the distance. "Hard to believe we've killed the last of them. I keep expecting someone to tap me on the shoulder and remind me that there are more unaccounted for." She paused, then mumbled, almost as if she were talking to herself: "we'll need to get George and his team up here for a final necropsy, somehow."

She looked down and shook her head. "This colony's been through hell, and it's not over. We've eliminated the primary threat, but we have a lot of work to do."

She turned to her surviving staff, and her voice cracked slightly. "I wish I could offer you a break, but…the best I can do is a few hours' sleep. We'll need to meet in the morning and start planning the colony's recovery. That will start with an assessment of what, and who, we've lost."

She sighed, "but, for now? Go get some sleep."

She turned and started walking, carefully, in the low gravity, back towards the access portal.

EPILOGUE

EPILOGUE

▪DAY 372

Elke leaned back in her chair, holding a shot of Irish whiskey. She was composing a welcoming message for their first incoming colony ship: 1,000 new colonists joining their ranks, bringing skills, enthusiasm, and livestock.

They would be in range to receive messages in just a few hours, so she needed to decide what else to say. She sighed, and as the chair reclined, she tilted her head back and stared at the ceiling. The chair had been a surprise gift, custom made for her a few months ago by Moira's team.

They'd given it to her after the Deep Space Service decided to make her the permanent colonial governor, in acknowledgment of her performance through the Kraken Crisis. She shook her head. *The Kraken Crisis...they had already given the ordeal a name.* It both humored her and disgusted her. She had been sure that they would replace her rather than keep her. Like most high achievers, she had difficulty accepting that she'd performed well; instead, she focused her status reports on what she should have done better. The Deep Space Service, however, reached a very different assessment based on the status reports from her senior staff and colonial leaders.

She sighed as she settled into the chair. She'd never considered a custom chair; getting used to it had taken time. Now, she felt foolish. She'd never realized how poorly she fit standard seating. It was *so* comfortable and well worth the space it consumed in her quarters.

Her quarters, now, *their* quarters. She smiled. Once her position as permanent colonial governor was confirmed, Elke and John decided to be more open about their relationship. The fact that nobody was surprised had, in fact, surprised *her*, much to John's amusement.

He was in Sick Bay at the moment, both to check on his patients and to give her some space to compose her message. He was still dealing with a reduced staff. The few that had been off duty when Sick Bay was attacked were supplemented by survivors cross-trained to perform some basic nursing services. They desperately needed more fully trained and experienced medical personnel; thankfully, several would be among the incoming colonists. Fortunately, they'd been spared any large scale events that would overwhelm their surviving medical personnel.

She found herself remembering Governor Kelty's welcome message as the *Lucky Strike* approached his colony. She recalled his description of Eden, the excitement about the opportunity for discovery. It had been a full year since their arrival. The first weeks, filled with marvelous exploration, seemed a distant memory; those weeks paled in comparison to their few days of horror and following months of challenging work.

The excitement and discovery had, of course, resumed. Eden remained a beautiful world filled with wonders, but their naivete was gone. Eden demanded respect, and was intolerant of mistakes. The inbound vessel would be here in a few days after a fifteen year journey (less than three for those on board). It would be the first of many. They knew nothing of the colony's first and terrible year. They would be filled with excitement and hope. She needed to inform them of the harsh lessons her people had learned, without dampening the enthusiasm of the new arrivals. It was a terrible balance to strike.

She thought about the hunt for the last dragon (as the diploid offspring were now called), and the search for missing people. Very, very few of the missing were found. The final toll, 143 souls, was a loss that still weighed heavily on the colony.

From a purely practical perspective, the loss of 143 from their initial complement was devastating. A number of key, skilled workers were lost, especially in Sick Bay. Survivors stepped up, willing to work long hours towards their recovery. All colonists volunteered for aptitude assessments and many were retrained until their small community was rebalanced and functioning again. All recovery costs were funded by the colony through the Jenna Faulk Memorial Fund.

But people were not merely practical considerations. Relatives and friends could never be replaced. There were no survivors that were not reeling from difficult losses. She felt this herself; she still missed Dani, terribly. Many of the survivors suffered lasting physical and emotional trauma. Jian was still recovering from her experience in Sick Bay; it was the same amongst the farmers. She was making progress. Preparing and publishing her reports on the kraken and dragons was therapeutic for her, but Elke wasn't sure whether she'd ever fully recover.

Many of the survivors, especially those that participated in the hunts, suffered long term brain damage from excessive electromagnetic suppression. Elke's headaches persisted for months, even as regenerative therapies restored her brain tissues. She was fortunate that she did not develop any cancers; a few unlucky victims did later on. Fortunately, John's Sick Bay was able to address and repair these conditions as they occurred.

And humans, being the adaptable creatures that they are, healed and progressed. For some, at least, life moved on. Survivors supported each other, and sometimes, found solace with each other. Families merged, pregnancies were announced. The first round of post-kraken babies had just recently been born, and many more were on the way.

As to the colony itself, Moira's prediction was correct. A few spores landed on vulnerable polymers, leading to small pinholes. A few of these polymers compromised externally facing systems, but these leaks were readily detected and repaired. A few compromised polymers weakened various systems supporting colonial operations and a very few affected industrial systems. Finding and repairing those problems was a bit more challenging than repairing structural air leaks, but they managed. After a few weeks, they began to believe that no more would plague them.

John's scans caught and readily treated a few early infections. After a few weeks without new patients, John recommended suspending weekly scans. All colonists were scanned at three-month and six-month intervals without further incident. Eventually, John planned one more pass of scans at the one year mark.

Crop damage was minimal, and Elke had been assured that the losses would be inconsequential, especially given the reduction in the colonial population. Sadly, fewer colonists meant fewer mouths to feed. Livestock was another story. Despite being bred with high reproductive and fast growth rates, the goat, pig, and chicken populations were taking what felt like far too long to recover. Milk and eggs were still rationed, and the only animal products available to colonists were surplus males and unproductive hens. The fabricated meats from the algal and insect farms were nutritious, but a poor substitute. She smiled. The vegetarians didn't mind; their diet didn't change very much. And there was fish, of course. But even fish had to be rationed, given the reduction in alternatives.

They were all looking forward to the inbound vessel's livestock: more goats, pigs, chickens, their first sheep and turkeys, more catfish, and additional varieties of tilapia. The first cattle would be on ships already heading their way; they could expect them in two or three years. They were still years away from being able to farm other aquatic staples, like trout.

The inbound ship would also have bees. Crops hadn't been affected much other than the destruction of bee colonies by the kraken. They didn't try to feed on the bees; they just destroyed the colonies as they found them. Something about the bees annoyed them. Perhaps it had something to do with their electromagnetic senses? Elke shrugged. It was now an academic point. They'd done their best to salvage the destroyed beehives to keep the crops pollinated, and the remaining honey had been rationed.

They'd also been rationing alcohol. As could be expected, there was a significant surge in alcohol consumption in the weeks after the kraken infestation. The surge, though it subsided, led to a shortage of anything not produced by the colony. As their stock of wine and spirits dwindled, there was a resurgence of moonshine.

Elke looked forward to putting an end to rationing. She shook her head. She's lost count of the number of times black markets emerged for alcohol and animal products. People would never change.

The landing parties resumed once an alternate polymer was available. It was no more or less resilient than the material they'd been using, but it was chemically distinct from anything they'd found on Eden. In lab tests, it tested impervious to spores. Still, they instituted pressure tests on return to the colony in addition to departure, and they randomly scanned landing party personnel. There were no new infestations to date. Walking on the surface without an environmental suit was, of course, not an option.

On the other hand, George's team discovered a few Eden species that were not susceptible to the spores, and they were still researching how these creatures resisted infestation. If they could find a natural substance that prevented the spores from implanting parasites, walking on Eden's surface might be an option again.

But, as George once explained, even if they could walk on the surface without environmental suits, farms would have to be under domes in order to purge the soils of native microorganisms. It was the only way to establish an Earthly ecosystem to support edible crops. Given the planned cadence of inbound colonial ships, orbiting platform expansion into the plenum was underway and plans were being drafted for the first dome and space elevator.

The first domed city would be named after Daniella Wu. Elke thought that a fitting memorial. Various parts of that city would also be named for other colonists lost; there were, unfortunately, plenty of names to memorialize.

The tsunami site succession was unfortunately nearly complete by the time George's team resumed their studies. But, Eden didn't keep them waiting long. Another marine quake led to another tsunami that wiped out an equatorial coastline in the southern hemisphere. George's team set up blinds immediately, and, this time, were able to observe kraken in their natural environment. They emerged from their hosts and hunted other pioneers, competing with predators that traveled to the site, and doing their best to evade a birdlike creature that specialized in preying on them. They were looking forward to seeing the diploid dragons emerge as the succession continued.

The ore deposit David and poor Esteban found in the northern hemisphere had proved rich with precious metals. They'd finished their drill cores and modeled the ore body; Cecilia was planning an extraction project that reflected what they were learning about ecological succession in temperate riparian Eden ecosystems. David and Cecilia's mining operations continued around the system; processing pods were ready to be fitted to the inbound deep space vessel frame for its return journey. The inbound vessel's residential and agricultural rings would be held in orbit and refit as processing pods for the next inbound vessel's return voyage. Elke smiled, thinking about David and Cecilia. They were a delightful couple, and she'd had the privilege of officiating when they formalized their relationship.

David was particularly excited about a recent find in one of the outer planets: a naturally occurring, stable, superheavy element. That could have enormous economic potential, if they can find it in quantity, and if Cecilia can develop an extraction model. So far, the only stable superheavy elements were produced in high energy physics labs. They had valuable properties, but were prohibitively expensive to produce. Naturally occurring deposits could bring even greater wealth to the colony. Time would tell…

She sighed, stood up, and set aside her empty glass. She stepped to the communications console in her quarters.

She was dressed, as always, in her casual uniform. She knew many of the colonists had adopted current fashions as reported through entertainment videos. But, after the kraken adventure, she decided to stick with her casual

Deep Space uniform. But she wore it without any rank insignia, something that was accepted by the colony as a more than tolerable quirk.

She said, "Begin" and, looking into the lens, smiled.

"Captain, welcome to Eden. I'm Governor Lubandi. We have some news and updates now that you've decelerated enough to receive messages."

■DAY 6127

Elke sat in the colonial bar, nursing a shot of Irish whiskey, looking out the window towards the surface of Eden. It was a very special whiskey; it was from the first batch produced by the first distillery on Eden. They used locally grown unmalted barley and aged it in a stout barrel. She sipped gently, and held the liquid in her mouth to savor it before swallowing. She nodded to herself as she contemplated the glass; they were off to an excellent start!

It was late in the day, and the sun was setting over the dome below. The lights of the space elevator came alive against the growing dark of the vast desert beneath them. The elevator stretched from the base of her orbiting colonial drum to the dome below, glowing against the looming night.

The population of the colonial drum had grown with each year's incoming colonists and the fecundity of her citizens. The thriving population was approaching 20,000. The plenum spaces, once converted to living and working space to accommodate a rapidly growing population, were now empty again as most of the colony moved to the surface. Colonial government remained in orbit, servicing interplanetary and interstellar traffic, along with applicable degree programs of the colony's Deep Space Academy.

Eden's Deep Space Academy was, understandably, best known for the finest xenobiological degree programs in known space; it had a long waiting list of students hoping to immigrate and enroll. Research labs had been established around the planet, funded by the enormous wealth extracted from Eden and the surrounding system of precious metals, superheavy elements, and organic chemicals. Compounds found in venoms, toxic plants, and other sources fueled significant advances in the pharmacology and polymers labs of the Deep Space Academy.

She finished her whiskey and put the glass down on the bar. It was a beautiful bar, the first she'd ever seen with real, albeit alien, wood.

She missed George. He'd moved to the dome to pursue field work after retiring as chair of the xenobiology program at the university. His pursuits especially focused on ecological succession. He worked extensively with the kraken in their natural habitats. He'd managed to prove his speculations about their biology during those dark days, so many years ago. She stayed abreast of his publications, and managed to connect with him occasionally; but usually, he was out of contact, roaming the planet's ecological disturbances.

She also missed David and Cecilia. They spent all of their time overseeing extraction projects on Eden and the asteroid belt. Nikko did, finally, settle down to start a family, finding happiness with a widowed colonist. He absolutely adored his two stepchildren, and together they added five more. Moira, after a few years, felt the itch, the deep space itch. She had signed on as chief engineer for one of the deep space vessels taking their exports to other systems. Elke looked forward to hearing from her in a few years when they arrived.

Her reverie was interrupted by a ping from her wrist computer. A glance informed her that she should take the message in her office. She put down the glass, and signaled to the bartender to put the whiskey on her tab. She walked through the corridor from the premium bar and restaurant territory of the ring to the governmental offices, finally arriving at her office. It adjoined her conference room and the new, spacious quarters she shared with the colony's chief medic.

She sat at a small conference table and flicked her wrist computer. Governor Gulati, who'd been appointed governor when Abram Kelty returned to the Deep Space Service, appeared before her.

"Governor Lubandi, I hope this communique finds you well. I know you've been waiting to hear the fate of the *Lucky Strike*. She arrived on schedule, and an automated message was received once she sufficiently decelerated."

Hearing of an automated message gave Elke a sense of foreboding.

"I'll forward the message to you, once you've confirmed receipt of this message. It was sent by Captain Rizzo, programmed to be forwarded on their arrival. He reported that a member of his crew was lost to an alien parasitic infestation, and that the parasites, after consuming the remains of their host and exhausting whatever vermin they could find, began to hunt the rest of his crew. He recorded the message when it became clear that it was a matter of time before they all fell prey to the things.

We dispatched a salvage team. The ship was largely intact, and none of the processing pods were affected. Some segments of the command ring were open to space; we believe they attempted to eliminate the predators through vacuum purges. Other segments were compromised by pinholes; they apparently, slowly, lost air.

He sighed, then continued. "No survivors were found. An alien skeleton, a presumed kraken skeleton based on your reports, was found. The team also found the skeletal remains of what we assume to be the crewmember whose initial infection led to the infestation of kraken."

He paused again. "I'm...so sorry to relay this news. While it is not pleasant news, I hope that it, at least, brings some closure to you and the members of your staff that knew the crew of the *Lucky Strike*. The processing pods will be delivered, and the balance of their sale will be credited to your colony's accounts. The ship will be restored and refitted for her next journey. In addition to Captain Rizzo's final message, we'll be forwarding a complete report of the salvage operation."

The projection ceased.

She leaned back and collected herself. She'd had fifteen years to prepare herself for this news, yet it still hurt. After she wiped away some loose tears from her cheeks, Elke began to prepare for a memorial service, in memory of the captain and crew of the *Lucky Strike*.

■ ■ ■

The captain of the salvage crew strode into the colonial bar. He could see, through the lateral observation windows, his interplanetary craft at a space buoy alongside the *Lucky Strike*. They'd towed the vessel back, and it was waiting for its refit.

They'd just docked, and, per tradition, they all headed to the bar. The five members of his crew walked in with him. Normally, after a profitable salvage operation, they'd be boisterous, talking about how they planned to spend their shares of the proceeds. But the scenes aboard the *Lucky Strike* were disturbing, and all were subdued.

The captain ordered a first round in memory of the late fellow travelers they'd never met. They all raised their glasses in toast, then swiftly drank their shots. After a solemn pause, the captain ordered another round. To his left, his navigator lifted his glass, took a sip, and resumed absentmindedly scratching, on his thigh, a red spot with a tiny black center.

ACKNOWLEDGMENTS

I'm grateful to my family for putting up with me locked away in my office for so many hours of writing and rewriting (and, of course, re-re-writing).

In particular, I want to thank my wife for encouraging me along the way. I also want to thank my sons for consultations on the writing process and giving feedback regarding the story, and my friend and dive buddy/mentor, Chris, for reviewing my speculations about diving on an alien world. It would have been very embarrassing, as a dive instructor, to mess that up.

This is my first attempt at a novel, so the work included not only the writing, but also the discovery of my personal writing process for fiction. I've written magazine articles, technical material (white papers and patent applications, never intentionally fictional), presentations, etc., but never before a novel.

The folks at Paperclip Publishing were fabulous to work with, particularly my editor, Abigail. In addition to catching my typos and corralling my run-on-sentences, she made numerous suggestions that made it a better book.

Hours and hours were spent working through the underlying science for "Kraken of Eden." I worked through what a realistic interstellar human expansion might entail, what might motivate human commitment to such expansion; I also had to consider what conditions might result in the evolution of a predator with sufficient digestive plasticity as to threaten our heroes, storyboarding the parallel stories of the kraken lifecycle, the exploration of Eden, the colonists' growing understanding of the danger in their midst, etc.

I hope you've enjoyed reading the story as much as I've enjoyed writing it. Thank you!

ABOUT THE AUTHOR

George Moakley started his career studying biology, with dreams of doing fieldwork and ecosystem modeling. To make ends meet, he took a data entry position with a precious metals company. Throughout his long career in the tech industry, George gained an inside perspective on various scientific and business practices, including designing Edge Intelligence solutions and conducting strategic planning workshops. He owns several patents, and holds prestigious advisory positions with ASU's School of Business and UCI's Customer Experience Program.

A fan of science fiction since childhood, George remains fascinated by the hard sciences. He feeds his passion for nature through photography while traveling the world; he also regularly partakes in hiking and scuba diving. Such travels have brought great joy, but also great concern about the fragility of our ecosystems. He has witnessed devastation caused by climate change, invasive species, overfishing, and other environmental issues. George is a long time member of the Nature Conservancy; a meeting with a conservancy representative in 2019 inspired a strategic thought experiment, regarding how the twenty-first century is likely to play out. The results were sobering, but provided the real-life inspiration behind Kraken of Eden.

Today, George makes his home in sunny Arizona with his inspiration, Diana. Between them, they have seven kids. When he's not locked away writing, he loves to visit his children and grandchildren, and can often be spotting driving back and forth from various sports activities.

Printed in the USA
CPSIA information can be obtained
at www.ICGtesting.com
LVHW010758290824
789620LV00001B/7

9 781734 620795